Main–Group Elements

18 VIII A

| | | | | | | | | | | | 2 4.002602 He helium |
| | | | 13 III A | 14 IV A | 15 V A | 16 VI A | 17 VII A | | | | |

			5 10.811 B boron	6 12.011 C carbon	7 14.00674 N nitrogen	8 15.9994 O oxygen	9 18.9984032 F fluorine	10 20.1797 Ne neon
10	11 I B	12 II B	13 26.981539 Al aluminum	14 28.0855 Si silicon	15 30.973762 P phosphorus	16 32.066 S sulfur	17 35.4527 Cl chlorine	18 39.948 Ar argon
28 58.69 Ni nickel	29 63.546 Cu copper	30 65.39 Zn zinc	31 69.723 Ga gallium	32 72.61 Ge germanium	33 74.92159 As arsenic	34 78.96 Se selenium	35 79.904 Br bromine	36 83.80 Kr krypton
46 106.42 Pd palladium	47 107.8682 Ag silver	48 112.411 Cd cadmium	49 114.82 In indium	50 118.710 Sn tin	51 121.75 Sb antimony	52 127.60 Te tellurium	53 126.90447 I iodine	54 131.29 Xe xenon
78 195.08 Pt platinum	79 196.96654 Au gold	80 200.59 Hg mercury	81 204.3833 Tl thallium	82 207.2 Pb lead	83 208.98037 Bi bismuth	84 (209) Po polonium	85 (210) At astatine	86 (222) Rn radon
110 (269) Uun ununnilium	111 (272) Uuu unununium	112 (277) Uub ununbiium						

Inner–Transition Metals

63 151.965 Eu europium	64 157.25 Gd gadolinium	65 158.92534 Tb terbium	66 162.50 Dy dysprosium	67 164.93032 Ho holmium	68 167.26 Er erbium	69 168.93421 Tm thulium	70 173.04 Yb ytterbium	71 174.967 Lu lutetium
95 (243) Am americium	96 (247) Cm curium	97 (247) Bk berkelium	98 (251) Cf californium	99 (252) Es einsteinium	100 (257) Fm fermium	101 (258) Md mendelevium	102 (259) No nobelium	103 (262) Lr lawrencium

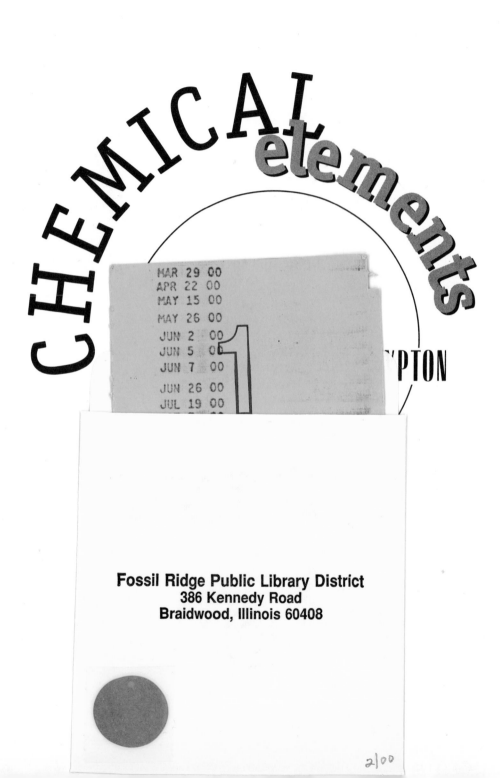

CHEMICAL elements

KRYPTON

MAR 29 00
APR 22 00
MAY 15 00
MAY 26 00
JUN 2 00
JUN 5 00
JUN 7 00
JUN 26 00
JUL 19 00

1

2/00

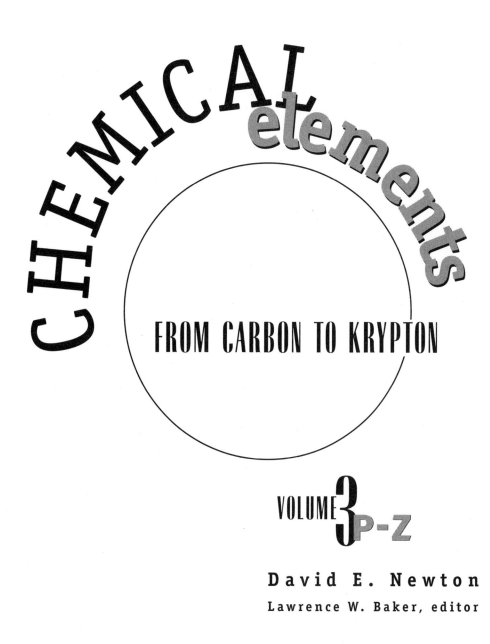

CHEMICAL elements

FROM CARBON TO KRYPTON

VOLUME **3** P-Z

David E. Newton

Lawrence W. Baker, editor

AN IMPRINT OF GALE

DETROIT · LONDON

Dedicated to three very special women in my life,
Agnes Greenhaw, Donna Krouse-Slesnick, and Ruth Ranen
"Thanks" is not enough for all you have done for me!

Chemical Elements: From Carbon to Krypton
David E. Newton

Staff

Lawrence W. Baker, *U•X•L Senior Editor*
Carol DeKane Nagel, *U•X•L Managing Editor*
Thomas L. Romig, *U•X•L Publisher*

Shalice Shah-Caldwell, *Permissions Associate (Pictures)*
Jessica L. Ulrich, *Permissions Associate (Pictures)*
Margaret Chamberlain, *Permissions Specialist (Pictures)*

Deborah Milliken, *Production Assistant*
Evi Seoud, *Assistant Production Manager*
Mary Beth Trimper, *Production Director*

Eric Johnson, *Art Director*
Cynthia Baldwin, *Product Design Manager*

Pamela Reed, *Photography Coordinator*
Mike Logusz, *Imaging Specialist*
Randy A. Bassett, *Image Database Supervisor*
Barbara J. Yarrow, *Graphic Services Manager*

Marco Di Vita, The Graphix Group, *Typesetting*

Library of Congress Cataloging-in-Publication Data

Newton, David E.
    Chemical elements : from carbon to krypton / David E. Newton, Lawrence W. Baker, editor.
        p.   cm.
    Includes bibliographical references and index.
    Contents: v. 1. A-F -- v. 2. G-O -- v. 3. P-Z.
    ISBN 0-7876-2844-1 (set). -- ISBN 0-7876-2845-X (v. 1). -- ISBN 0-7876-2846-8 (v. 2). -- ISBN 0-7876-2847-6 (v. 3)
    1. Chemical elements. I. Baker, Lawrence W.
QD466.N464 1998
546--dc21   98-31207
                                                                                  CIP

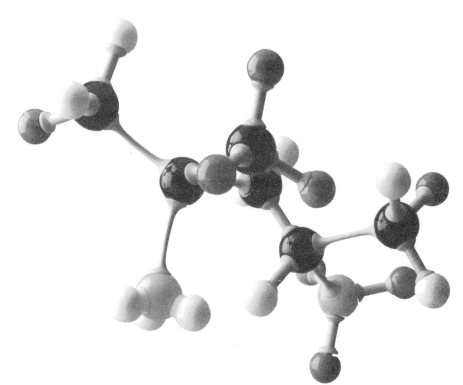

CONTENTS

Volume 1

Volume 2

Volume 3

CHEMICAL elements

Contents

CHEMICAL**elements**

CONTENTS
Elements by Atomic Number

Bold-Italic type indicates volume numbers

CHEMICAL elements

CHEMICAL **elements**

CHEMICAL **elements**

CONTENTS

Elements by Family Group

Bold-Italic type indicates
volume numbers

Group 10 (VIIIB)

Group 11 (IB)

Group 12 (IIB)

Group 13 (IIIA)

Group 14 (IVA)

Group 15 (VA)

Group 16 (VIA)

Group 17 (VIIA)

Group 18 (VIIIA)

Lanthanides

Actinides

Contents: Elements by Family Group

READER'S GUIDE

Many young people like to play with Lego blocks, tinker-toys, erector sets, and similar building games. It's fun to see how many different ways a few simple shapes can be put together.

The same can be said of chemistry. The world is filled with an untold number of different objects, ranging from crystals and snowflakes to plant and animal cells to plastics and medicines. Yet all of those objects are made from various combinations of only about 100 basic materials: the chemical elements.

Scientists have been intrigued about the idea of an "element" for more than two thousand years. The early Greeks developed complicated schemes that explained everything in nature using only a few basic materials, such as earth, air, fire, and water. The Greeks were wrong in terms of the materials they believed to be "elemental." But they were on the right track in developing the concept that such materials did exist.

By the 1600s, chemists were just beginning to develop a modern definition of an element. An element, they said, was any object that cannot be reduced to some simpler form of matter. Over the next 300 years, research showed that about 100 such materials exist. These materials range from such well known elements as oxygen, hydrogen, iron, gold, and silver to substances that are not at all well known, elements such as neodymium, terbium, rhenium, and seaborgium.

By the mid-1800s, the search for new chemical elements had created a new problem. About 50 elements were known at the time. But no one yet knew how these different elements related to each other, if they did at all. Then, in one of the great coincidences in chemical history, that question was answered independently by two scientists at almost the same time, German chemist Lothar Meyer and Russian chemist Dmitri Mendeleev. (Meyer, however, did not publish his research until 1870, nor did he predict the existence of undiscovered elements as Mendeleev did.)

Meyer and Mendeleev discovered that the elements could be grouped together to make them easier to study. The grouping occurred naturally when the elements were laid out in order, according to their atomic weight. Atomic weight is a quantity indicating atomic mass that tells how much matter there is in an element or how dense it is. The product of Meyer and Mendeleev's research is one of the most famous visual aids in all of science, the periodic table. Nearly every classroom has a copy of this table. It lists all of the known chemical elements, arranged in rows and columns. The elements that lie within a single column or a single row all have characteristics that relate to each other. Chemists and students of chemistry use the periodic table to better understand individual elements and the way the elements are similar to and different from each other.

About *Chemical Elements: From Carbon to Krypton*

Chemical Elements: From Carbon to Krypton is designed as an introduction to the chemical elements. Elements with atomic numbers 1 through 100 are examined in separate entries, while the transfermium elements (elements 101 through 112) are covered in one entry.

Students will find *Chemical Elements* useful in a number of ways. First, it is a valuable source of fundamental information for research reports, science fair projects, classroom demonstrations, and other activities. Second, it can be used to provide more detail about elements and compounds that are only mentioned in other science textbooks or classrooms. Third, it will be an interesting source of information about the building blocks of nature for those who simply want to know more about the elements.

The three-volume set is arranged alphabetically by element name. Each entry contains basic information about the element discussed: its discovery and naming, physical and chemical properties, isotopes, occurrence in nature, methods of extraction, important compounds and uses, and health effects.

The first page of each entry provides basic information about the chemical element: its chemical symbol, atomic number, atomic mass, family, and pronunciation. A diagram of an atom of the element is also shown, with the atom's electrons arranged in various "energy levels" outside the nucleus. Inside the nucleus, the number of protons and neutrons is indicated.

Entries are easy to read and written in a straightforward style. Difficult words are defined within the text. Each entry also includes a "Words to Know" section that defines technical words and scientific terms. This enables students to learn vocabulary appropriate to chemistry without having to consult other sources for definitions.

Added features

Chemical Elements: From Carbon to Krypton includes a number of additional features that help make the connection between elements, minerals, the people who discovered and worked with them, and common uses of the elements.

- Three tables of contents: alphabetically by element name; by atomic number; and by family group provide varied access to the elements.

- A timeline at the beginning of each volume provides a chronology of the discovery of the elements.

- Nearly 200 photographs and illustrations of the elements and products in which they are used bring the elements to life.

- Sidebars provide fascinating supplemental information about scientists, theories, uses of elements, and more.

- Interesting facts about the elements are highlighted in the margins.

- Extensive cross references make it easy to read about related elements. Other elements mentioned within an element's entry are boldfaced upon first mention, serving as a helpful reminder that separate entries are written for these elements.

- A list of sources for further reading for some elements and for general chemistry is found at the end of each volume.

- A comprehensive index quickly points readers to the elements, minerals, and people mentioned in *Chemical Elements: From Carbon to Krypton*.

- A periodic table on the endsheets gives students a quick look at the elements.

Special thanks

The editor wishes to thank imaging guru Randy Bassett for his patience and guidance. Thanks also to Bernard Grunow for his informal assistance in the early stages of the editing phase. Kudos to typesetter Marco Di Vita, who, as always, is in a league by himself. And, finally, a big-time thank-you to soulmate Beth Baker, whose editorial toolbelt, no doubt, needs some duct tape by now.

Comments and suggestions

We welcome your comments on this work as well as suggestions for future science titles. Please write: Editors, *Chemical Elements: From Carbon to Krypton,* U•X•L, 27500 Drake Rd., Farmington Hills, Michigan, 48331-3535; call toll-free: 800-347-4253; send fax to 248-699-8066; or send e-mail via http://www.gale.com.

TIMELINE
The Discovery of Elements

Early history	The elements **carbon, sulfur, iron, tin, lead, copper, mercury, silver,** and **gold** are known to humans.
	Pre-a.d. 1600: The elements **arsenic, antimony, bismuth,** and **zinc** are known to humans.
1669	German physician Hennig Brand discovers **phosphorus.**
1735	Swedish chemist Georg Brandt discovers **cobalt.**

c. 4000 b.c.	c. 1200 b.c.	c. 6 b.c.	1492	1620
The Bronze Age begins	The Iron Age begins	Jesus of Nazareth is born	Christopher Columbus sails to the Americas	Pilgrims land at Plymouth, Mass.

| c. 4000 b.c. | c. 1200 b.c. | c. 6 b.c. | 1492 | 1620 |

c. 1748	Spanish military leader Don Antonio de Ulloa discovers **platinum.**
1751	Swedish mineralogist Axel Fredrik Cronstedt discovers **nickel.**
1766	English chemist and physicist Henry Cavendish discovers **hydrogen.**
1772	Scottish physician and chemist Daniel Rutherford discovers **nitrogen.**
1774	Swedish chemist Carl Wilhelm Scheele discovers **chlorine.**
1774	Swedish mineralogist Johann Gottlieb Gahn discovers **manganese.**
1774	English chemist Joseph Priestley and Swedish chemist Carl Wilhelm Scheele discover **oxygen.**
1781	Swedish chemist Peter Jacob Hjelm discovers **molybdenum.**
c. 1782	Austrian mineralogist Baron Franz Joseph Müller von Reichenstein discovers **tellurium.**
1783	Spanish scientists Don Fausto D'Elhuyard and Don Juan José D'Elhuyard, and Swedish chemist Carl Wilhelm Scheele discover **tungsten.**

1746
Benjamin Franklin experiments with electricity

1746

1775
American Revolution begins

1775

1789
French Revolution begins

1789

| 1789 | German chemist Martin Klaproth discovers **uranium.** |

1789 — German chemist Martin Klaproth discovers **uranium.**

1789 — German chemist Martin Klaproth discovers **zirconium.**

1791 — English clergyman William Gregor discovers **titanium.**

1794 — Finnish chemist Johan Gadolin discovers **yttrium.**

1797 — French chemist Louis-Nicolas Vauquelin discovers **chromium.**

1798 — French chemist Louis-Nicolas Vauquelin discovers **beryllium.**

1801 — English chemist Charles Hatchett discovers **niobium.**

1801 — Spanish-Mexican metallurgist Andrés Manuel del Río discovers **vanadium.**

1802 — Swedish chemist and mineralogist Anders Gustaf Ekeberg discovers **tantalum.**

1803 — English chemist and physicist William Hyde Wollaston discovers **palladium.**

1803 — Swedish chemists Jöns Jakob Berzelius and Wilhelm Hisinger, and German chemist Martin

1789 George Washington is elected first U.S. president

1782 *Farmer's Almanac* is first published

1794 Cotton gin is patented

1800 Washington, D.C. becomes U.S. capitol

Klaproth discover black rock of Bastnas, Sweden, which led to the discovery of several elements.

| 1804 | English chemist and physicist William Hyde Wollaston discovers **rhodium.** |

1804   English chemist and physicist William Hyde Wollaston discovers **rhodium.**

1804   English chemist Smithson Tennant discovers **osmium.**

1804   English chemist Smithson Tennant discovers **iridium.**

1807   English chemist Sir Humphry Davy discovers **potassium.**

1807   English chemist Sir Humphry Davy discovers **sodium.**

1808   English chemist Sir Humphry Davy discovers **barium.**

1808   English chemist Sir Humphry Davy discovers **strontium.**

1808   English chemist Sir Humphry Davy discovers **calcium.**

1808   English chemist Sir Humphry Davy discovers **magnesium.**

1804
Lewis & Clark
expedition begins

1806
Webster's Dictionary is
first published

1804

1806

1808	French chemists Louis Jacques Thênard and Joseph Louis Gay-Lussac discover **boron.**
1811	French chemist Bernard Courtois discovers **iodine.**
1817	Swedish chemist Johan August Arfwedson discovers **lithium.**
1817	German chemist Friedrich Stromeyer discovers **cadmium.**
1818	Swedish chemists Jöns Jakob Berzelius and J. G. Gahn discover **selenium.**
1823	Swedish chemist Jöns Jakob Berzelius discovers **silicon.**
1825	Danish chemist and physicist Hans Christian Oersted discovers **aluminum.**
1826	French chemist Antoine-Jérôme Balard discovers **bromine.**
1828	Swedish chemist Jöns Jakob Berzelius discovers **thorium.**
1830	Swedish chemist Nils Gabriel Sefström rediscovers **vanadium.**
1839	Swedish chemist Carl Gustav Mosander discovers **cerium.**

1808
Humphry Davy invents carbon arc lamp

1812
War of 1812 begins

1814
Francis Scott Key writes "Star Spangled Banner"

1837
Queen Victoria II begins reign over England

| 1808 | 1812 | 1814 | 1837 |

1839	Swedish chemist Carl Gustav Mosander discovers **lanthanum.**
1843	Swedish chemist Carl Gustav Mosander discovers **terbium.**
1843	Swedish chemist Carl Gustav Mosander discovers **erbium.**
1844	Russian chemist Carl Ernst Claus discovers **ruthenium.**
c. 1861	German chemists Robert Bunsen and Gustav Kirchhoff discovers **cesium.**
c. 1861	German chemists Robert Bunsen and Gustav Kirchhoff discovers **rubidium.**
1861	British physicist Sir William Crookes discovers **thallium.**
1863	German chemists Ferdinand Reich and Hieronymus Theodor Richter discovers **indium.**
1875	Paul-émile Lecoq de Boisbaudran discovers **gallium.**
1878	Jean-Charles-Galissard de Marignac receives partial credit for the discovery of **ytterbium.**
1879	Swedish chemist Per Teodor Cleve discovers **holmium.**

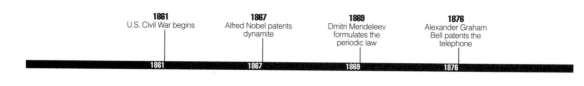

1861	**1867**	**1869**	**1876**
U.S. Civil War begins	Alfred Nobel patents dynamite	Dmitri Mendeleev formulates the periodic law	Alexander Graham Bell patents the telephone
1861	1867	1869	1876

CHEMICAL **elements**

1879	Swedish chemist Per Teodor Cleve discovers **thulium.**
1879	Swedish chemist Lars Nilson discovers **scandium.**
1879	Swedish chemist Lars Nilson receives partial credit for the discovery of **ytterbium.**
1880	French chemist Paul-Émile Lecoq de Boisbaudran discovers **samarium.**
1880	French chemist Jean-Charles-Galissard de Marignac discovers **gadolinium.**
1885	Austrian chemist Carl Auer (Baron von Welsbach) discovers **praseodymium.**
1885	Austrian chemist Carl Auer (Baron von Welsbach) discovers **neodymium.**
1885	German chemist Clemens Alexander Winkler discovers **germanium.**
1886	French chemist Henri Moissan discovers **fluorine.**
1886	French chemist Paul-Émile Lecoq de Boisbaudran discovers **dysprosium.**
1894	English chemists Lord Rayleigh and William Ramsay discover **argon.**

1884
A worldwide system of standard time is adopted
1884

1888
George Eastman introduces the Kodak camera
1888

1892
Rudolf Diesel patents the internal-combustion engine
1892

1895	English chemist Sir William Ramsay and Swedish chemists Per Teodor Cleve and Nils Abraham Langlet discover **helium.**
1898	English chemists William Ramsay and Morris Travers discover **krypton.**
1898	English chemists William Ramsay and Morris Travers discover **neon.**
1898	English chemists William Ramsay and Morris Travers discover **xenon.**
1898	French physicists Marie and Pierre Curie discover **polonium.**
1898	French physicists Marie and Pierre Curie discover **radium.**
1899	French chemist André Debierne discovers **actinium.**
1900	German physicist Friedrich Ernst Dorn discovers **radon.**
1901	French chemist Eugène-Anatole Demarçay discovers **europium.**
1907	French chemist Georges Urbain discovers **lutetium.**

1896
Henry Ford assembles the first motor car

1898
Marie and Pierre Curie publish papers on radioactivity

1903
Wilbur and Orville fly first plane at Kitty Hawk, N.C.

1905
Albert Einstein formulates the theory of relativity

1896   1898   1903   1905

1907	French chemist Georges Urbain receives partial credit for the discovery of **ytterbium.**
1917	German physicists Lise Meitner and Otto Hahn discover **protactinium.**
1923	Dutch physicist Dirk Coster and Hungarian chemist George Charles de Hevesy discover **hafnium.**
1925	German chemists Walter Noddack, Ida Tacke, and Otto Berg discover **rhenium.**
1933	French chemist Marguerite Perey discovers **francium.**
1939	Italian physicist Emilio Segrè and his colleague Carlo Perrier discover **technetium.**
1940	Edwin M. McMillan (1907–91) and Philip H. Abelson prepare **neptunium.**
1940	Dale R. Corson, Kenneth R. Mackenzie, and Emilio Segrè discover **astatine.**
1940	University of California at Berkeley researcher Glenn Seaborg and others prepare **plutonium.**
1944	University of California at Berkeley researchers Glenn Seaborg, Albert Ghiorso, Ralph A.

Timeline: The Discovery of Elements

1912
The *Titanic* hits an iceberg and sinks

1914
World War I begins

1926
Robert Goddard launches the first liquid-propellant rocket

1929
Great Depression begins

1939
World War II begins

1912   1914   1926   1929   1939

James, and Leon O. Morgan prepare **americium.**

1944   University of California at Berkeley researchers Glenn Seaborg, Albert Ghiorso, and Ralph A. James prepare **curium.**

1945   Scientists at the Oak Ridge Laboratory in Oak Ridge, Tennessee, discover **promethium.**

1949   University of California at Berkeley researchers prepare **berkelium.**

1950   University of California at Berkeley researchers Glenn Seaborg, Albert Ghiorso, Kenneth Street, Jr., and Stanley G. Thompson prepare **californium.**

1954   University of California at Berkeley researchers prepare **einsteinium.**

1954   University of California at Berkeley researcher Albert Ghiorso and others prepare **fermium.**

1960s & 1970s Researchers at the Joint Institute of Nuclear Research, in Dubna, Russia; the Lawrence Berkeley Laboratory at the University of California at Berkeley; and the Institute for Heavy Ion Research in Darmstadt, Germany, continue to prepare new transfermium elements.

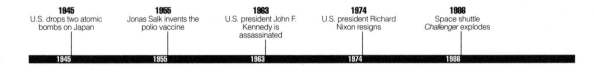

1945	**1955**	**1963**	**1974**	**1986**
U.S. drops two atomic bombs on Japan	Jonas Salk invents the polio vaccine	U.S. president John F. Kennedy is assassinated	U.S. president Richard Nixon resigns	Space shuttle *Challenger* explodes
1945	1955	1963	1974	1986

WORDS TO KNOW

A

Abrasive a powdery material used to grind or polish other materials

Absolute zero the lowest temperature possible, about −273°C (−459°F)

Actinide family elements in the periodic table with atomic numbers 90 through 103

Alchemy a kind of pre-science that existed from about 500 B.C. to about the end of the 16th century

Alkali metal an element in Group 1 (IA) of the periodic table

Alkali a chemical with properties opposite those of an acid

Alkaline earth metal an element found in Group 2 (IIA) of the periodic table

Allotropes forms of an element with different physical and chemical properties

Alloy a mixture of two or more metals that has properties different from those of the individual metals

Alpha particles tiny, atom-sized particles that can destroy cells

Alpha radiation a form of radiation that consists of very fast moving alpha particles and helium atoms without their electrons

Amalgam a combination of mercury and at least one other metal

Amorphous without crystalline shape

Anhydrous ammonia dry ammonia gas

Antiseptic a chemical that stops the growth of germs

Aqua regia a mixture of hydrochloric and nitric acids that often reacts with materials that do not react with either acid separately

B

Battery a device for changing chemical energy into electrical energy

Biochemistry the field of chemistry concerned with the study of compounds found in living organisms

Biocompatible not causing a reaction when placed into the body

Bipolar disorder a condition in which a person experiences wild mood swings

Brass an alloy of copper and zinc

Bronze Age a period in human history ranging from about 3500 B.C. to 1000 B.C., when bronze was widely used for weapons, utensils, and ornamental objects

Bronze an alloy of copper and tin

Buckminsterfullerene full name for buckyball or fullerene; *see* Buckyball

Buckyball an allotrope of carbon whose 60 carbon atoms are arranged in a sphere-like form

C

Capacitor an electrical device, somewhat like a battery, that collects and then stores up electrical charges

Carat a unit of weight for gold and other precious metals, equal to one fifth of a gram, or 200 milligrams

Carbon arc lamp a lamp for producing very bright white light

Carbon-14 dating a technique that allows archaeologists to estimate the age of once-living materials by using the knowledge that carbon-14 is found in all living carbon materials; once an organism dies, no more carbon-14 remains

Cassiterite an ore of tin containing tin oxide, the major commercial source of tin metal

Catalyst a substance used to speed up or slow down a chemical reaction without undergoing any change itself

Chalcogens elements in Group 16 (VIA) of the periodic table

Chemical reagent a substance, such as an acid or an alkali, used to study other substances

Chlorofluorocarbons (CFCs) a family of chemical compounds consisting of carbon, fluorine, and chlorine that were once used widely as propellants in commercial sprays but regulated in the United States since 1987 because of their harmful environmental effects

Corrosive agent a material that tends to vigorously react or eat away at something

Cyclotron a particle accelerator, or "atom smasher," in which small particles, such as protons, are made to travel very fast and then collide with atoms, causing the atoms to break apart

D

Density the mass of a substance per unit volume

Diagnosis finding out what medical problems a person may have

Distillation a process by which two or more liquids can be separated from each other by heating them to their boiling points

"Doped" containing a small amount of a material as an impurity

Ductile capable of being drawn into thin wires

E

Earth in mineralogy, a naturally occurring form of an element, often an oxide of the element

Electrolysis a process by which a compound is broken down by passing an electric current through it

Electroplating the process by which a thin layer of one metal is laid down on top of a second metal

Enzyme a substance that stimulates certain chemical reactions in the body

F

Fabrication shaping, molding, bending, cutting, and working with a metal

Fission the process by which large atoms break apart, releasing large amounts of energy, smaller atoms, and neutrons in the process

Fly ash the powdery material produced during the production of iron or some other metal

Frasch method a method for removing sulfur from underground mines by pumping hot air and water down a set of pipes

Fuel cell any system that uses chemical reactions to produce electricity

Fullerene alternative name for buckyball; *see* Buckyball

G

Galvanizing the process of laying down a thin layer of zinc on the surface of a second metal

Gamma rays a form of radiation similar to X rays

H

Half life the time it takes for half of a sample of a radioactive element to break down

Halogen one of the elements in Group 17 (VIIA) of the periodic table

Heat exchange medium a material that picks up heat in one place and carries it to another place

Hydrocarbons compounds made of carbon and hydrogen

Hypoallergenic not causing an allergic reaction

I

Inactive does not react with any other element

Inert gases *see* **Noble gases**

Inert not very active

Isotope two or more forms of an element that differ from each other according to their mass number

L

Lanthanide family the elements in the periodic table with atomic numbers 58 through 71

Laser a device for making very intense light of one very specific color that is intensified many times over

Liquid air air that has been cooled to a very low temperature

Luminescence the property of giving off light without giving off heat

CHEMICAL **elements**

M

Machining the bending, cutting, and shaping of a metal by mechanical means

"Magic number" the number of protons and/or neutrons in an atom that tend to make the atom stable (not radioactive)

Magnetic field the space around an electric current or a magnet in which a magnetic force can be observed

Malleable capable of being hammered into thin sheets

Metals elements that have a shiny surface, are good conductors of heat and electricity, and can be melted, hammered into thin sheets, and drawn into thin wires

Metalloid an element that has characteristics of both metals and non-metals

Metallurgy the art and science of working with metals

Micronutrient a substance needed in very small amounts to maintain good health

Misch metal a metal that contains different rare earth elements and has the unusual property of giving off a spark when struck

Mohs scale a way of expressing the hardness of a material

Mordant a material that helps a dye stick to cloth

N

Nanotubes long, thin, and extremely tiny tubes

Native not combined with any other element

Neutron radiography a technique that uses neutrons to study the internal composition of material

Nickel allergy a health condition caused by exposure to nickel metal

Nitrogen fixation the process of converting nitrogen as an element to a compound that contains nitrogen

Noble gases elements in Group 18 (VIIIA) of the periodic table

Noble metals see **Platinum family**

Non-metals elements that do not have the properties of metals

Nuclear fission a process in which neutrons collide with the nucleus of a plutonium or uranium atom, causing it to split apart with the release of very large amounts of energy

Nuclear reactor a device in which nuclear reactions occur

O

Optical fiber a thin strand of glass through which light passes; the light carries a message, much as an electric current carries a message through a telephone wire

Ore a mineral compound that is mined for one of the elements it contains, usually a metal element

Organic chemistry the study of the carbon compounds

Oxidizing agent a chemical substance that gives up or takes on electrons from another substance

Ozone a form of oxygen that filters out harmful radiation from the sun

P

Particle accelerator ("atom smasher") a device used to cause small particles, such as protons, to move at very high speeds

Periodic law a way of organizing the chemical elements to show how they are related to each other

Periodic table a chart that shows how chemical elements are related to each other

Phosphor a material that gives off light when struck by electrons

Photosynthesis the process by which plants convert carbon dioxide and water to carbohydrates (starches and sugars)

Platinum family a group of elements that occur close to platinum in the periodic table and with platinum in the Earth's surface

Polymerization the process by which many thousands of individual tetrafluoroethlylene (TFE) molecules join together to make one very large molecule

Potash a potassium compound that forms when wood burns

Precious metal a metal that is rare, desirable, and, therefore, expensive

Proteins compounds that are vital to the building and growth of cells

Pyrophoric gives off sparks when scratched

Q

Quarry a large hole in the ground from which useful minerals are taken

R

Radiation energy transmitted in the form of electromagnetic waves or subatomic particles

Radioactive isotope an isotope that breaks apart and gives off some form of radiation

Radioactive tracer an isotope whose movement in the body can be followed because of the radiation it gives off

Radioactivity the process by which an isotope or element breaks down and gives off some form of radiation

Rare earth elements *see* **Lanthanide family**

Reactive combines with other substances relatively easily

Refractory a material that can withstand very high temperatures and reflects heat back away from itself

Rodenticide a poison used to kill rats and mice

Rusting a process by which a metal combines with oxygen

S

Salt dome a large mass of salt found underground

Semiconductor a material that conducts an electric current, but not nearly as well as metals

Silver plating a process by which a very thin layer of silver metal is laid down on top of another metal

Slag a mixture of materials that separates from a metal during its purification and floats on top of the molten metal

Slurry a soup-like mixture of crushed ore and water

Solder an alloy that can be melted and then used to join two metals to each other

Spectra the lines produced when chemical elements are heated

Spectroscope A device for analyzing the light produced when an element is heated

Spectroscopy the process of analyzing light produced when an element is heated

Spectrum (plural: spectra) the pattern of light given off by a glowing object, such as a star

Stable not likely to react with other materials

Sublimation the process by which a solid changes directly to a gas when heated, without first changing to a liquid

Superalloy an alloy made of iron, cobalt, or nickel that has special properties, such as the ability to withstand high temperatures and attack by oxygen

Superconductivity the tendency of an electric current to flow through a material without resistance

Superconductor a material that has no resistance to the flow of electricity; once an electrical current begins flowing in the material, it continues to flow forever

Superheated water water that is hotter than its boiling point, but which has not started to boil

Surface tension a property of liquids that makes them act like they are covered with a skin

T

Tarnishing oxidizing; reacting with oxygen in the air

Tensile capable of being stretched without breaking

Thermocouple a device for measuring very high temperatures

Tin cry a screeching-like sound made when tin metal is bent

Tin disease a change that takes place in materials containing tin when the material is cooled to temperatures below 13°C for long periods of time, when solid tin turns to a crumbly powder

Tincture a solution made by dissolving a substance in alcohol

Tinplate a type of metal consisting of thin protective coating of tin deposited on the outer surface of some other metal

Toxic poisonous

Trace element an element that is needed in very small amounts for the proper growth of a plant or animal

Tracer a radioactive isotope whose presence in a system can easily be detected

Transfermium element any element with an atomic number greater than 100

Transistor a device used to control the flow of electricity in a circuit

Transition metal an element in Groups 3 through 12 of the periodic table

Transuranium element an element with an atomic number greater than 92

U

Ultraviolet (UV) radiation electromagnetic radiation (energy) of a wavelength just shorter than the violet (shortest wavelength) end of the visible light spectrum and thus with higher energy than visible light

V
Vulcanizing the process by which soft rubber is converted to a harder, longer-lasting product

W
Workability the ability to work with a metal to get it into a desired shape or thickness

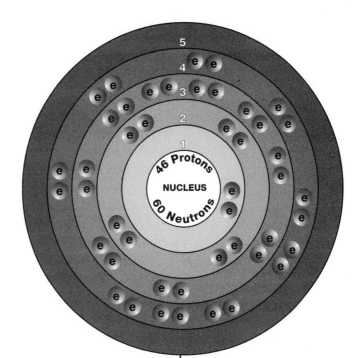

46 Protons
NUCLEUS
60 Neutrons

PALLADIUM

Overview

Palladium is found in Row 5, Group 10 (VIIIB) of the periodic table. The periodic table is a chart that shows how chemical elements are related to each other. Palladium, **ruthenium, rhodium, osmium, iridium,** and **platinum** make up the platinum group of metals. These metals are also sometimes called the noble metals. That term reflects the fact that the six elements are not very reactive.

Palladium was discovered along with rhodium in 1803 by English chemist William Hyde Wollaston (1766–1828). Wollaston had been studying platinum ores, probably taken from South America.

Like the other platinum metals, palladium is quite rare. It also has a beautiful shiny finish that does not tarnish easily. These properties make it desirable in making jewelry and art objects. These uses are its most important applications.

Discovery and naming

As early as the 1700s, Brazilian miners described a number of platinum-like metals. They called them by such names as *prata* (silver), *ouro podre* (worthless or spoiled gold), and *ouro bran-*

SYMBOL
Pd

ATOMIC NUMBER
46

ATOMIC MASS
106.42

FAMILY
Group 10 (VIIIB)
Transition metal;
platinum group

PRONUNCIATION
puh-LAY-dee-um

co (white gold). These names may or may not have matched the materials that were actually present in the metals. For example, a substance the miners called *platino* (platinum) was probably a combination of **gold** and palladium.

In the early 1800s, Wollaston received samples of some of these metals. He decided to analyze them. During his work, he found that a sample of platinum ore contained other metals as well. These metals turned out to be two new elements—rhodium and palladium. The name palladium was taken from Pallas, an asteroid that had been discovered at about the same time.

Chemists later found palladium in other South American ores. A sample of ouro podre, for example, turned out to include about 86 percent gold, 10 percent palladium, and 4 percent silver.

Physical properties

Palladium is a soft, silver-white metal. It is both malleable and ductile. Malleable means capable of being hammered into thin sheets. Ductile means capable of being drawn into thin wires. The malleability of palladium is similar to that of gold. It can be hammered into sheets no more than about a millionth of a centimeter thick.

An interesting property of palladium is its ability to absorb (soak up) hydrogen gas like a sponge. When a surface is coated with finely divided palladium metal, the hydrogen gas passes into the space between palladium atoms. Palladium absorbs up to 900 times its own weight in hydrogen gas.

Chemical properties

Palladium has been called "the least noble" of the noble metals because it is the most reactive of the platinum group. It combines poorly with **oxygen** under normal conditions but will catch fire if ground into powder. Palladium does not react with most acids at room temperature but will do so when mixed with most hot acids. The metal will also combine with **fluorine** and **chlorine** when very hot.

Occurrence in nature

The abundance of palladium in the Earth's crust is estimated to be about 1 to 10 parts per trillion. That makes it one of the ten rarest elements found in the Earth's crust. It usually occurs

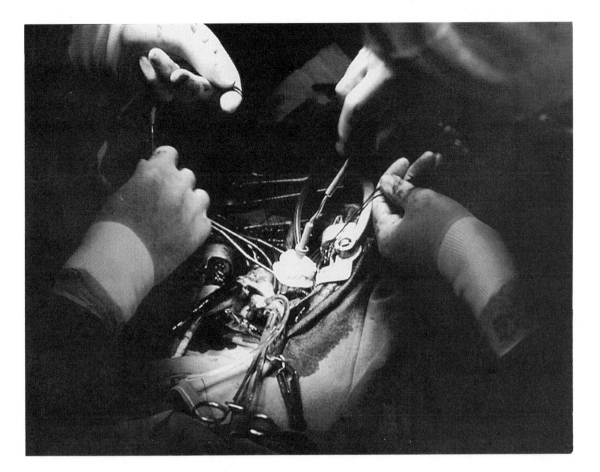

in native form, meaning "not combined with any other element." Palladium is usually found with platinum and other members of the noble metal group.

Palladium alloys are commonly used in surgical instruments, such as those used here in open-heart surgery.

Russia and South Africa produce about 93 percent of the palladium mined in the world. It is also found in Canada and the United States.

Palladium combines poorly with oxygen under normal conditions but will catch fire if ground into powder.

Isotopes

There are six naturally occurring isotopes of palladium: palladium-102, palladium-105, palladium-106, palladium-108, and palladium-110. Isotopes are two or more forms of an element. Isotopes differ from each other according to their mass number. The number written to the right of the element's name is the mass number. The mass number represents the number of protons plus neutrons in the nucleus of an atom of the element. The number of protons determines the element, but the

Palladium alloys are used to make ball bearings. They are shown here in a castor, the part of a wheel that enables a chair to swivel and roll.

number of neutrons in the atom of any one element can vary. Each variation is an isotope.

About a dozen radioactive isotopes of palladium are known also. A radioactive isotope is one that breaks apart and gives off some form of radiation. Radioactive isotopes are produced when very small particles are fired at atoms. These particles stick in the atoms and make them radioactive.

No isotope of palladium has an important commercial use.

Extraction

Palladium is removed from platinum ores after platinum and gold have been removed. The metal is converted to palladium chloride ($PdCl_2$) and then purified as pure palladium.

Uses

Palladium has two primary uses: as a catalyst and in making jewelry and specialized alloys. A catalyst is a substance used to speed up or slow down a chemical reaction without undergoing any change itself. Palladium catalysts are used in breaking down petroleum to make high quality gasoline and other products. It is also used in the production of some essential chemicals, such as sulfuric acid (H_2SO_4), which is used in paper and fabric production. The catalytic convertors used in automobiles today may also contain a palladium catalyst. A catalytic convertor is a device added to a car's exhaust system. It helps the fuel used in the car burn more efficiently.

An alloy is made by mixing two or more melted metals. The solid mixture has properties different from those of the individual metals. Palladium is commonly alloyed with gold, **silver,** and **copper.** The alloys are used in a variety of products, such as ball bearings, springs, balance wheels of watches, surgical instruments, electrical contacts, and astronomical mirrors. Palladium alloys are also used quite widely in dentistry.

Compounds

Relatively few palladium compounds are commercially important.

Health effects

There is no evidence of serious health effects from exposure to palladium or its compounds.

Many automobiles use a palladium catalyst to help fuel burn more efficiently.

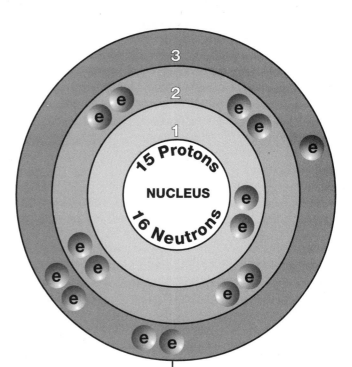

PHOSPHORUS

Overview

Phosphorus is found in Group 15 (VA) of the periodic table. The periodic table is a chart that that shows how chemical elements are related to each other. Phosphorus is part of the nitrogen family along with **nitrogen, arsenic, antimony,** and **bismuth.**

Phosphorus was first discovered in 1669 by German physician Hennig Brand (ca. 1630–92). Brand is somewhat famous in chemistry. He is sometimes called the last of the alchemists. Alchemy was a kind of pre-science that existed from about 500 B.C. to about the end of the 16th century. Alchemists wanted to find a way of changing **lead, iron,** and other metals into **gold.** They also wanted to find a way of having eternal life. Alchemy contained too much magic and mysticism to be a real science. But it developed a number of techniques and produced many new materials that were later found to be useful in modern chemistry.

Brand was convinced that the key to changing metals into gold could be found in urine. He decided to look for the "magic substance" that could change lead into gold in urine. In the process of heating and purifying urine, he obtained phospho-

SYMBOL
P

ATOMIC NUMBER
15

ATOMIC MASS
30.97376

FAMILY
Group 15 (VA)
Nitrogen

PRONUNCIATION
FOS-fer-us

Alchemy a kind of pre-science that existed from about 500 B.C. to about the end of the 16th century.

Allotropes forms of an element with different physical and chemical properties

Catalyst a substance used to speed up or slow down a chemical reaction without undergoing any change itself

Halogen one of the elements in Group 17 (VIIA) of the periodic table

Isotopes two or more forms of an element that differ from each other according to their mass number

Periodic table a chart that shows how chemical elements are related to each other

Radioactive isotope an isotope that breaks apart and gives off some form of radiation

Sublimation the process by which a solid changes directly to a gas when heated, without first changing to a liquid

Tracer a radioactive isotope whose presence in a system can easily be detected

rus. The discovery was important because it was the first time someone had discovered an element not known to ancient peoples. In that regard, Brand was the first person who could be called the discoverer of an element.

Phosphorus is a fascinating element that occurs in at least three very different forms. If left exposed to the air, it catches fire on its own. It also glows in the dark. Today, its most important use is in the manufacture of phosphoric acid (H_3PO_4). Phosphoric acid, in turn, is used to manufacture fertilizers and a number of other less important products.

Discovery and naming

Phosphorus and its compounds may have been known before Brand's discovery. Old manuscripts refer to materials that glow in the dark. The word used for such materials today is phosphorescent. Early Christians noted the use of "perpetual lamps" that glowed in the dark. The lamps may have contained phosphorus or one of its compounds.

Still, Brand was the first to record the process of making pure phosphorus. No one knows how he decided that urine might contain a chemical that could be used to turn lead into gold. His experiments to find such a chemical were, of course, a failure. But he made an accidental discovery along the way. That discovery was a material that glowed in the dark: phosphorus.

Scientists were fascinated when they heard of Brand's discovery. They tried to repeat his research. Some tried to talk him into selling his discovery to kings and princes. The new element seemed to be a way of getting rich and becoming famous.

But Brand was never interested in these ideas. Instead, he gave away all of the phosphorus he prepared. Other scientists soon began to experiment with the element. One of the first discoveries they made was how dangerous phosphorus is. One scientist wrote that a servant left some phosphorus on top of his bed one day. Later that night, the bed covers burst into flame. The phosphorus had caught fire by itself!

Eventually, Brand's method of making phosphorus became widely known. The element joined iron, gold, silver, arsenic, and the handful of other elements known to early chemists.

Little is known about what happened to Brand after his discovery. In fact, there is no record of where or when he died.

Physical properties

Phosphorus exists in at least three allotropic forms. Allotropes are forms of an element with different physical and chemical properties. The three main allotropes are named for their colors: white phosphorus (also called yellow phosphorus), red phosphorus, and black phosphorus (also called violet phosphorus). These allotropes all have different physical and chemical properties.

White phosphorus is a waxy, transparent solid. Its melting point is 44.1°C (111°F) and its boiling point is 280°C (536°F). It has a density of 1.88 grams per cubic centimeter. If kept in a vacuum, it sublimes if exposed to light. Sublimation is the process by which a solid changes directly to a gas when heated, without first changing to a liquid. White phosphorus is phosphorescent. It gives off a beautiful greenish-white glow. It does not dissolve well in water, although it does dissolve in other liquids, such as benzene, chloroform, and **carbon** disulfide. White phosphorus sometimes appears slightly yellowish because of traces of red phosphorus.

Red phosphorus is a red powder. It can be made by heating white phosphorus with a catalyst to 240°C (464°F). A catalyst is a substance used to speed up or slow down a chemical reaction without undergoing any change itself. Without a catalyst, red phosphorus sublimes at 416°C (781°F). Its density is 2.34 grams per cubic centimeter. It does not dissolve in most liquids.

Black phosphorus looks like graphite powder. Graphite is a form of carbon used in "lead" pencils. Black phosphorus can be made by applying extreme pressure to white phosphorus. It has a density of 3.56 to 3.83 grams per cubic centimeter. One of its interesting properties is that it conducts an electric current in spite of being a non-metal.

Chemical properties

White phosphorus is the form that occurs most commonly at room temperatures. It is very reactive. It combines with **oxygen** so easily that it catches fire spontaneously (automatically). As a safety precaution, white phosphorus is stored under water in chemical laboratories.

Brand was convinced that the key to changing metals into gold could be found in urine. Instead, he found phosphorus.

Dry and wet red phosphorus solids.

Phosphorus combines easily with the halogens. The halogens are the elements that make up Group 17 (VIIA) of the periodic table. They include **fluorine, chlorine, bromine, iodine,** and **astatine.** For example, the reaction between phosphorus and chlorine is:

$$2P + 3Cl_2 \rightarrow 2PCl_3$$

Phosphorus also combines with metals to form compounds known as phosphides:

$$3Mg + 2P \rightarrow Mg_3P_2 \text{ (magnesium phosphide)}$$

Occurrence in nature

The abundance of phosphorus in the Earth's crust is estimated to be 0.12 percent, making it the 11th most common element. It usually occurs as a phosphate. A phosphate is a compound that contains phosphorus, oxygen, and at least one more element. An example is calcium phosphate, $Ca_3(PO_4)_2$.

White phosphorus combines with oxygen so easily that it catches fire automatically. As a safety precaution, white phosphorus is stored under water in chemical laboratories.

The only important commercial source of phosphorus is phosphate rock. Phosphate rock is primarily calcium phosphate. The United States is the largest producer of phosphate rock in the world. In 1996, 13,300,000 metric tons of phosphate rock were mined in the United States. That amounted to about a third of the world's total phosphate rock.

About 86 percent of phosphate rock comes from North Carolina and Florida. Smaller amounts are also mined in Idaho and Utah. Other major producers of phosphate rock are Morocco, China, Russia, Tunisia, Jordan, and Israel.

Isotopes
Only one naturally occurring isotope of phosphorus exists, phosphorus-31. Isotopes are two or more forms of an element. Isotopes differ from each other according to their mass number. The number written to the right of the element's name is the mass number. The mass number represents the number of protons plus neutrons in the nucleus of an atom of the element. The number of protons determines the element, but the number of neutrons in the atom of any one element can vary. Each variation is an isotope.

Six radioactive isotopes of phosphorus are known also. A radioactive isotope is one that breaks apart and gives off some form of radiation. Radioactive isotopes are produced when very small particles are fired at atoms. These particles stick in the atoms and make them radioactive.

One radioactive isotope, phosphorus-32, has applications in medicine, industry, and tracer studies. A tracer is a radioactive isotope whose presence in a system can easily be detected. The isotope is injected into the system where it gives off radiation. The radiation is followed by means of detectors placed around the system.

Phosphorus-32 is especially useful in medical studies, because phosphorus occurs in many parts of the body. Radioactive phosphorus can be used as a tracer to study parts of the body as well as chemical changes inside the body. Radioactive phosphorus can also determine how much blood is in a person's body. It can also help locate the presence of tumors in the brain, eyes, breasts, and skin. Finally, it is sometimes used to treat certain

Phosphorus

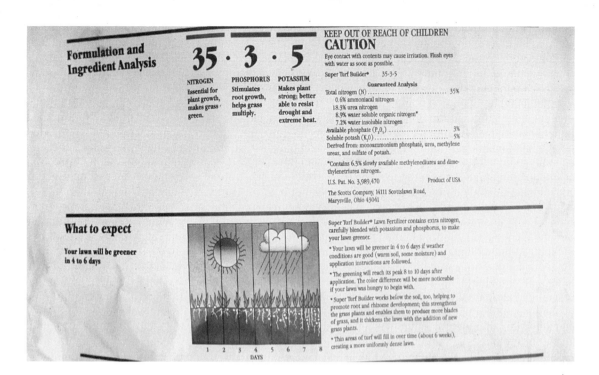

A bag of fertilizer shows information about its nitrogen, phosphorus, and potassium ingredients.

Radioactive phosphorus helps locate the presence of tumors in the brain, eyes, breasts, and skin.

forms of cancer. Radiation given off by the phosphorus-32 may kill cancer cells and help slow or stop the disease.

Phosphorus-32 is important in a variety of scientific studies. For example, it is added to tires when they are made. Then, the radiation it gives off can be studied as the tires are used. This information tells where the tire wears out and how long it takes to wear out.

Extraction

It is possible to make pure phosphorus from phosphate rock. The rock is mixed with sand and coke (pure carbon). The mixture is then heated in an electric furnace. An electric furnace is a device for producing very high temperatures. Pure phosphorus is produced in this reaction. It escapes from the mixture as a vapor (gas). The cooled vapor solidifies into white phosphorus. The reaction is:

$$Ca_3(PO_4)_2 + 3SiO_2 + 5C \rightarrow 3CaSiO_3 + 5CO + 2P$$

This reaction is not very important because pure phosphorus has few uses. The most important compounds of phosphorus are all made from phosphate rock or calcium phosphate. Therefore, the most important step in producing "phosphorus" is

Two phosphorus compounds are used to make the coating found on the tips of safety matches.

simply to separate pure calcium phosphate from phosphate rock. This can be done fairly easily.

Uses and compounds

In 1996, 91 percent of all the phosphate rock mined in the United States was used to make fertilizer. Modern farmers use enormous amounts of synthetic (artificial) fertilizer on their crops. This synthetic fertilizer contains nitrogen, phosphorus, and **potassium,** the three elements critical to growing plants. These elements normally occur in the soil, but may not be present in large enough amounts. Adding them by means of synthetic fertilizer helps plants grow better. Most farmers add some form of synthetic fertilizer to their fields every year. This demand for synthetic fertilizers accounts for the major use of phosphorus compounds.

Phosphorus and its compounds have other uses. These uses account for about 10 percent of all the phosphorus produced. For example, the compounds known as phosphorus pentasulfide (P_2S_5) and phosphorus sesquisulfide (P_4S_3) are used to make ordinary wood and paper safety matches. These compounds coat the tip of the match. When the match is scratched on a surface, the phosphorus pentasulfide or phosphorus sesquisulfide bursts into flame. It ignites other chemicals on the head of the match.

Two phosphorus compounds are used to coat the tip of a match.

Another compound of phosphorus with a number of uses is phosphorus oxychloride ($POCl_3$). This compound is used in the manufacture of gasoline additives, in the production of certain kinds of plastics, as a fire retardant agent, and in the manufacture of transistors for electronic devices.

Health effects

Phosphorus is essential to the health of plants and animals. Many essential chemicals in living cells contain phosphorus. One of the most important of these chemicals is adenosine triphosphate (ATP). ATP provides the energy to cells they need to stay alive and carry out all the tasks they have to perform. Phosphorus is critical to the development of bones and teeth. Nucleic acids also contain phosphorus. Nucleic acids are chemicals that perform many functions in living organisms. For example, they carry the genetic information in a cell. They tell the cell what chemicals it must make. It also acts as the "director" in the formation of those chemicals.

The daily recommended amount of phosphorus for humans is one gram. It is fairly easy to get that much phosphorus every day through meat, milk, beans, and grains.

On the other hand, elemental phosphorus is extremely dangerous. Elemental phosphorus is phosphorus as an element, not combined with other elements. Swallowing even a speck of white phosphorus produces severe diarrhea with loss of blood; damage to the liver, stomach, intestines, and circulatory sys-

Making lakes *too* healthy

The second most important use of phosphate compounds is in making detergents. The compound most often used in detergents is called sodium tripolyphosphate, or STPP ($Na_5P_3O_{10}$).

STPP adds a number of benefits to a detergent. For example, it can kill some bacteria and prevent washers from becoming corroded (rusted) and clogged. The most important function in detergents, however, is as a water-softening agent.

Natural water often contains chemicals that keep soaps and detergents from sudsing. They reduce the ability of soaps and detergents to clean clothes. STPP has the ability to capture these chemicals. It greatly improves the ability of soaps and detergents to make suds and clean clothes. The first detergent to use STPP was Tide, in 1947. The introduction of Tide brought about a revolution in clothes cleaning.

But STPP can create problems for the environment. After detergents have been used, they often end up in rivers and streams and, eventually, in lakes from waste water. And that's just fine for the algae that live in those lakes. Algae are tiny green plants that use phosphorus as they grow. As more detergents get into lakes, the amount of STPP increases. That means there is more phosphorus in a lake and that, in turn, means that algae begin to grow much faster.

In some cases, there is so much STPP and phosphorus in a lake that algae grow out of control, clogging the lake with algae and other green plants. The lake slowly turns into a swamp, and finally into a meadow. The lake disappears!

Many people became concerned about this problem in the 1960s. They demanded that less STPP be used in detergents. A number of cities and states banned the sale of STPP detergents. STPP production had grown rapidly from 1.10 billion pounds in 1955 to 2.44 billion pounds in 1970. But then production began to drop off. By the mid-1990s, production had dropped well below a billion pounds a year.

tem (blood flow system); and coma. Swallowing a piece of white phosphorus no larger than 50 to 100 milligrams (0.0035 ounce) can even cause death.

Handling white phosphorus is dangerous as well. It causes serious skin burns.

Interestingly, red phosphorus does not have the same effects. It is considered to be relatively safe. It is dangerous only if it contains white phosphorus mixed with it.

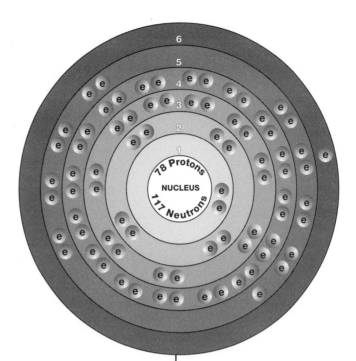

PLATINUM

Overview

Platinum is a transition metal in Group 10 (VIIIB) of the periodic table. The periodic table is a chart that shows how chemical elements are related to each other. Platinum is also a member of a group of metals named after itself. Other platinum metals include **ruthenium, rhodium, palladium, osmium, and iridium.** They are found in Rows 5 and 6 of Groups 8 through 10 in the periodic table. Platinum is also considered to be a precious metal. A precious metal is one that is rare and desirable.

The platinum group metals are sometimes referred to as the noble metals. That term comes from the fact that they are all relatively inactive. They do not combine with or interact with most other elements or compounds. This chemical inactivity accounts for some of the uses of the platinum metals. For example, platinum is often used to make laboratory equipment because it will not react with materials that come into contact with the equipment.

The primary use of platinum and other platinum metals is as catalysts. A catalyst is a substance used to speed up or slow down a chemical reaction without undergoing any change

SYMBOL
Pt

ATOMIC NUMBER
78

ATOMIC MASS
195.08

FAMILY
Group 10 (VIIIB)
Transition metal;
platinum group

PRONUNCIATION
PLAT-num

itself. For example, the catalytic convertor in an automobile's exhaust system may contain a platinum metal.

Discovery and naming

The first known reference to platinum can be found in the writings of Italian physician, scholar, and poet Julius Caesar Scaliger (1484–1558). Scaliger apparently saw platinum while visiting Central America in 1557. He referred to a hard metal that the natives had learned to work with, but the Spanish had not. The metal had been called *platina* ("little silver") by the natives. The name was given to the material because it got in the way of mining **silver** and **gold.** Since the natives knew of no use for the platina, they thought of it as a nuisance.

The first complete description of platinum was given by the Spanish military leader Don Antonio de Ulloa (1716–95). While serving in South America from 1735 to 1746, de Ulloa collected samples of platinum. He later wrote a report about the metal, describing how it was mined and used. De Ulloa is often given credit for discovering platinum on the basis of the report he wrote.

Reports of the new element spread through Europe. Scientists were fascinated by its physical properties. It was not only beautiful to look at, but resistant to corrosion (rusting). Many people saw that it could be used in jewelry and art objects, as with gold and silver. Demand for the metal began to grow, leading to what was then called the "Platinum Age in Spain."

Physical properties

Platinum is a silver-gray, shiny metal that is both malleable and ductile. Malleable means capable of being hammered into thin sheets. Platinum can be hammered into a fine sheet no more than 100 atoms thick, thinner than aluminum foil.

Ductile means the metal can be drawn into thin wires. Platinum has a melting point of about 1,773°C (3,223°F) and a boiling point of about 3,827°C (6,921°F). Its density is 21.45 grams per cubic centimeter, making it one of the densest elements.

Chemical properties

Platinum is a relatively inactive metal. When exposed to air, it does not tarnish or corrode. It is not attacked by most acids, but will dissolve in aqua regia. Aqua regia is a mixture of

WORDS TO KNOW

Alloy a mixture of two or more metals with properties different from those of the individual metals

Catalyst a substance used to speed up or slow down a chemical reaction without undergoing any change itself

Ductile capable of being drawn into thin wires

Isotopes two or more forms of an element that differ from each other according to their mass number

Malleable capable of being hammered into thin sheets

Periodic table a chart that shows how the chemical elements are related to each other

Radioactive isotope an isotope that breaks apart and gives off some form of radiation

hydrochloric and nitric acids. It often reacts with materials that do not react with either acid separately. Platinum also dissolves in very hot alkalis. An alkali is a chemical with properties opposite those of an acid. Sodium hydroxide ("common lye") and limewater are examples of alkalis.

An unusual property of platinum is that it will absorb large quantities of **hydrogen** gas at high temperatures. The platinum soaks up hydrogen the way a sponge soaks up water.

Occurrence in nature
The platinum metals are often found together in nature. In fact, one of the problems in producing platinum is finding a way of separating it from the other platinum metals. Unlike gold, however, these metals do not occur in masses large enough to mine. Instead, they are usually obtained as by-products from mining other metals, such as **copper** and **nickel.**

Platinum is one of the rarest elements. Its abundance is estimated to be about 0.01 parts per million in the Earth's crust. The world's largest supplier of platinum by far is South Africa. In 1996, that nation produced 117,000 kilograms of platinum. The next largest producer was Canada, producing only 8,260 kilograms in 1996. The only other large producer of platinum is the United States. Most of the platinum in the United States comes from the Stillwater Mine in Montana.

Isotopes

Six naturally occurring isotopes of platinum exist: platinum-190, platinum-192, platinum-194, platinum-195, platinum-196, and platinum-198. Of these, only platinum-190 is radioactive. Isotopes are two or more forms of an element. Isotopes differ from each other according to their mass number. The number written to the right of the element's name is the mass number. The mass number represents the number of protons plus neutrons in the nucleus of an atom of the element. The number of protons determines the element, but the number of neutrons in the atom of any one element can vary. Each variation is an isotope.

Artificially radioactive isotopes of platinum have also been produced. These isotopes are produced when very small particles are fired at atoms. These particles stick in the atoms and make them radioactive.

No radioactive isotope of platinum has any commercial application.

Extraction

The major challenge in obtaining pure platinum is separating it from other platinum metals. The first step in this process is to dissolve the mixture in aqua regia. Platinum dissolves in aqua regia, and other platinum metals do not. Platinum metal can then be removed from the aqua regia in a form known as platinum sponge. Platinum sponge is a sponge-like material of black platinum powder. Finally, the powder is heated to very high temperatures and melted to produce the pure metal.

Uses

If asked, most people would probably name jewelry as the most important use of platinum. And the metal *is* used for that purpose. It is hard, beautiful, corrosion-resistant—ideal for making bracelets, earrings, pins, watch bands, and other types of jewelry.

However, jewelry is not the most important use of platinum. The making of catalysts is. For example, platinum catalysts are

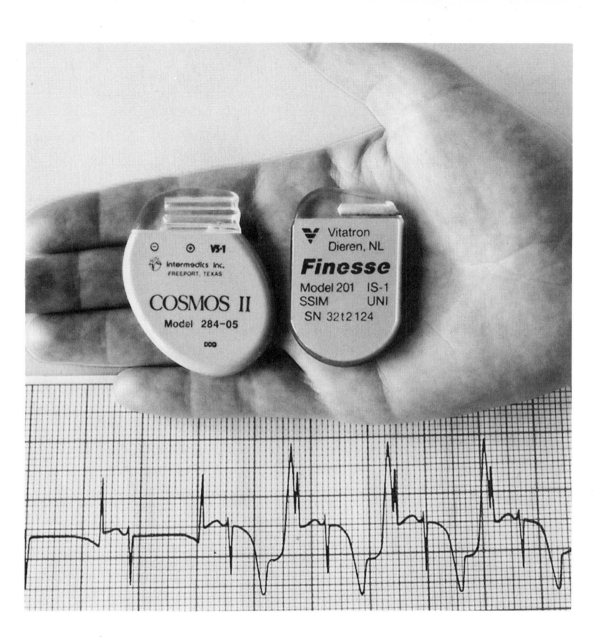

widely used in the modern petroleum industry. Crude oil from the ground must be treated before it can be converted to gasoline, fuel oil, and other petroleum products. The molecules must be broken apart, rearranged, and put back together again in new patterns. Platinum is one of the most important catalysts in making these reactions happen.

Platinum catalysts are also used to make compounds that end up as fertilizers, plastics, synthetic fibers, drugs and pharmaceuticals, and dozens of other everyday products. For example, platinum is used in the manufacture of nitric acid (HNO_3).

Artificial pacemakers are usually made out of platinum.

Nitric acid is used to produce ammonia, which, in turn, is used to make fertilizers.

Probably the best known use of platinum as a catalyst is in cars. All new automobiles have a catalytic convertor in the exhaust system. A catalytic convertor is a device that helps gasoline burn more completely. It reduces the amount of pollutants released to the air. Most catalytic convertors contain platinum or other platinum metals.

Platinum is used in other parts of a car or truck. Certain types of spark plugs, for example, may contain platinum. Overall, the greatest single use of platinum in the United States is in the manufacture of automobiles and trucks.

Many uses of platinum depend on its chemical inactivity. For example, some people have to have artificial heart pacemakers implanted into their chests. An artificial pacemaker is a device that makes sure the heart beats in a regular pattern. It usually replaces a body part that performs that function but has been damaged. Artificial pacemakers are usually made out of platinum. The platinum protects the pacemaker from corroding or being destroyed by acids inside the body.

Platinum is also used in small amounts in alloys. For example, **cobalt** alloyed with platinum makes a powerful magnet. An alloy is made by melting and mixing two or more metals. The mixture has properties different from those of the individual metals. The platinum-cobalt magnet is one of the strongest magnets known.

Compounds
Relatively few platinum compounds are commercially important.

Health effects
Platinum dust and some platinum compounds can have mild health effects. If inhaled, they can cause sneezing, irritation of the nose, and shortness of breath. If spilled on the skin, they can cause a rash and skin irritation.

Artificial pacemakers are usually made out of platinum. The platinum protects the pacemaker from corroding or being destroyed by acids inside the body.

PLUTONIUM

Overview

Plutonium is a synthetic (artificial) element. It exists naturally only in the smallest imaginable amounts. Plutonium was first prepared artificially by a team of researchers at the University of California at Berkeley (UCB) in 1941. News of this discovery was not released, however, until 1946. This delay was caused by the need for secrecy about scientific developments during World War II (1939–45).

Plutonium is a member of the actinide family. The actinides occur in Row 7 of the periodic table. The periodic table is a chart that shows how chemical elements are related to one another. The actinides get their name from element 89, **actinium,** which is sometimes considered the first member of the family. Plutonium is also called a transuranium element. The term transuranium means "beyond **uranium.**" Elements with atomic numbers greater than that of uranium (92) are called transuranium elements.

Plutonium has two important uses. First, some of its isotopes will undergo nuclear fission. Nuclear fission is a process in which an element is bombarded with neutrons. The element breaks apart into simpler elements, releasing large amounts of

SYMBOL
Pu

ATOMIC NUMBER
94

ATOMIC MASS
244.0642

FAMILY
Actinide
Transuranium element

PRONUNCIATION
plu-TOE-nee-um

energy. Plutonium has been used to make nuclear weapons (such as "atomic bombs") and in nuclear power plants to produce electricity. Plutonium has also been used as a portable energy supply in space probes and other space vehicles.

Discovery and naming

In 1940, American physicists Edwin McMillan (1907–91) and Philip Abelson (1913–) discovered the first transuranium element, **neptunium** (atomic number 93). The neptunium they produced was radioactive. They predicted it would break down to form a new element, atomic number 94. But McMillan and Abelson were called away to do research on the atomic bomb. They suggested to a colleague, Glenn Seaborg (1912–), that he continue their research on neptunium.

Seaborg and his associates picked up where McMillan and Abelson had left off. They eventually proved that element 94 did exist. The proof came in an experiment they conducted in a particle accelerator at UCB. A particle accelerator is sometimes called an "atom smasher." It is used to cause small particles, such as protons, to move at very high speeds. The particles then collide with targets, such as **gold, copper,** or **tin.** When struck by the particles, the targets break apart, forming new elements and other particles.

Seaborg's team suggested the name plutonium for the new element, in honor of the planet Pluto. The two elements just before plutonium in the periodic table had also been named for planets: uranium for Uranus and neptunium for Neptune.

Glenn Seaborg later went on to find a number of other elements. One of those elements, atomic number 106, has been named seaborgium in his honor. (See **transfermium elements** in this volume.)

Physical properties

Plutonium is a silvery-white metal with a melting point of 639.5°C (1,183°F) and a density of 19.816 grams per cubic centimeter.

Chemical properties

Plutonium is highly reactive and forms a number of different compounds.

WORDS TO KNOW

Actinide family elements with atomic numbers 90 through 103

Half life the time it takes for half of a sample of a radioactive element to break down

Isotopes two or more forms of an element that differ from each other according to their mass number

Nuclear fission a process in which neutrons collide with the nucleus of a uranium atom causing it to split apart with the release of very large amounts of energy

Periodic table a chart that shows how chemical elements are related to each other

Radioactive isotope an isotope that breaks apart and gives off some form of radiation

Opposite page:
Plutonium is used to make nuclear weapons. Here, a computer-enhanced image shows the mushroom-clouded aftermath of the dropping of the atomic bomb over Nagasaki, Japan, on August 9, 1945.

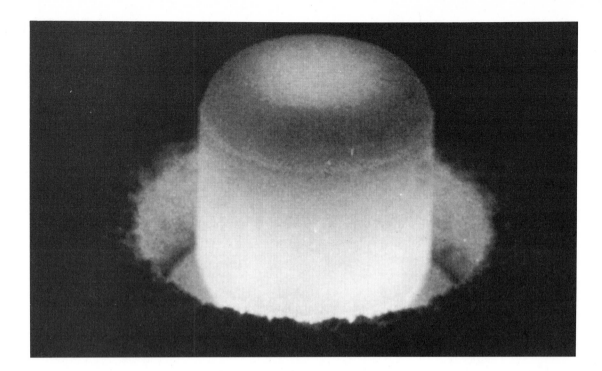

Glowing pellet of plutonium.

Occurrence in nature

Scientists now know that very small amounts of plutonium occur in the Earth's crust. It is formed in ores of uranium. When uranium breaks down, it sometimes forms plutonium in very small quantities. Scientists believe that the abundance of plutonium in the earth is about one quintillionth parts per million.

Isotopes

About 15 isotopes of plutonium are known to exist. All of these isotopes are radioactive. Isotopes are two or more forms of an element. Isotopes differ from each other according to their mass number. The number written to the right of the element's name is the mass number. The mass number represents the number of protons plus neutrons in the nucleus of an atom of the element. The number of protons determines the element, but the number of neutrons in the atom of any one element can vary. Each variation is an isotope.

The most stable isotopes of plutonium are plutonium-242 and plutonium-244. The half lives of these two isotopes are 376,300 years and 82,600,000 years respectively. The half life of a radioactive element is the time it takes for half of a sample of the element to break down. Consider the isotope pluto-

Plutonium is named after the planet Pluto.

Historic cardiac pacemaker fueled by radioactive plutonium-238.

nium-242, with its half life of 376,300 years. In 376,300 years (one half life), only half of a sample prepared today would still be plutonium-242. The rest would have broken down into a new isotope.

Extraction
Plutonium is extracted from natural sources only rarely and only for the purposes of research.

Uses
The most important uses of plutonium depend on two of its properties. First, the radiation given off by plutonium occurs as heat. In fact, plutonium gives off so much heat that the metal feels warm when it is touched. If a large piece of pluto-nium is placed into water, the heat released can cause the water to boil.

Plutonium provides electrical power on space probes and space vehicles.

This property makes plutonium a good choice for certain ther-moelectric generator applications. A thermoelectric generator is a device that converts heat into electricity. Plutonium gen-erators are not practical on a large scale basis. But they are very desirable for special conditions. For example, they have been used to provide electrical power on space probes and

Using fuel to make fuel

The production of plutonium fuel (plutonium-239) is a fascinating story. When nuclear reactors were first built, they all used uranium-235 as a fuel. Of the three naturally occurring isotopes of uranium, only uranium-235 will undergo fission.

But the uranium used in a nuclear reactor is never pure uranium. Instead, it is natural uranium with an increased amount of uranium-235. The uranium is said to be "enriched" with uranium-235. But a lot of the main isotope of uranium, uranium-238, remains mixed with the uranium-235.

Fission of uranium-235 occurs when neutrons are fired into the reactor. Neutrons are subatomic particles with no electric charge. They cause uranium-235 to break apart, giving off energy. That energy is then used to make electricity.

But neutrons also collide with uranium-238 isotopes in the reactor. This isotope does not undergo fission, but does undergo another kind of change. It soaks up neutrons and changes into plutonium-239:

uranium-238 + neutrons → plutonium-239

The plutonium that is formed can be removed from the reactor. It is then purified and re-used as fuel in another nuclear reactor.

What an amazing process this is! One could compare it to the burning of coal to make electricity. In a coal-fired power plant, coal is burned to boil water. Steam runs turbines that make electricity, but when the coal burns up, it's gone.

In a nuclear reactor, the breakdown of uranium-235 atoms gives off energy like the burning of coal. Over time, most of the uranium-235 atoms are used up. But while this is happening, a new fuel is being made! Atoms of plutonium-239 are being produced from atoms of uranium-238. Some reactors are operated primarily to make plutonium, not to make electricity. These reactors are called breeder reactors because they generate new fuel as they operate.

space vehicles. They have also been used in artificial pacemakers for people with heart conditions. The isotope most commonly used for this application is plutonium-238 because the radiation it gives off does not pose a threat to people's health.

Plutonium is also used as a fuel in nuclear power plants and in making nuclear weapons ("atomic bombs"). The isotope used for this purpose is plutonium-239. It is used because it will undergo nuclear fission. Very few isotopes will undergo nuclear fission. Two isotopes of uranium, uranium-233 and uranium-235, are among these. But uranium-233 does not occur at all in nature and uranium-235 occurs in only very small amounts.

By contrast, plutonium-239 can be made fairly easily in nuclear power reactors. It is a by-product, or "waste product," of these reactors. It can be removed from the reactor, purified, and then re-used to make electrical power.

Compounds
No plutonium compounds have any commercial application.

Health effects
Plutonium is one of the most toxic elements known. In the body, it tends to concentrate in bones. One of its most serious health effects on a long-term basis is bone cancer. Scientists who work with plutonium do not handle the metal directly. Instead, they use remote control devices. They always stand behind special shielding to protect themselves from the radiation produced by plutonium.

Plutonium is one of the most toxic elements known. Scientists do not handle it directly. They use remote control devices and stand behind special shielding to protect themselves from the radiation produced by plutonium.

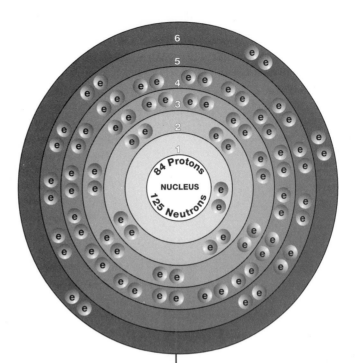

POLONIUM

Overview

Polonium was discovered in 1898 by Polish-French physicist Marie Curie (1867–1934) and her husband, Pierre (1859–1906). They were looking for the source of radioactivity in a naturally occurring ore called pitchblende. Radioactivity is the process by which materials give off energy and change into new materials.

Polonium is the heaviest element in the chalcogen family. It is in Group 16 (VIA) on the periodic table. The periodic table is a chart that shows how chemical elements are related to one another. The other chalcogen elements are **oxygen, sulfur, selenium,** and **tellurium.**

Polonium is a relatively rare element. The pitchblende studied by the Curies contained only about 100 micrograms (millionths of a gram) of polonium per metric ton of ore. The element can now be prepared artificially in a particle accelerator, or "atom smasher." It causes small particles such as protons, to move at very high speeds. These speeds approach the speed of light— 300,000,000 meters per second (186,000 miles per second). The particles collide with targets, usually **gold, copper,** or **tin.** When struck by the particles, the targets break apart, forming new elements and other particles.

SYMBOL
Po

ATOMIC NUMBER
84

ATOMIC MASS
208.9824

FAMILY
Group 16 (VIA)
Chalcogen

PRONUNCIATION
puh-LO-nee-um

Polonium has a few commercial uses. For example, it is used to remove static electrical charges in certain industrial operations. The element is highly toxic.

Discovery and naming

In 1898 French physicist Antoine-Henri Becquerel (1852–1908) had discovered a new form of radiation that was similar to light rays. It was found in a **uranium** ore called pitchblende.

Becquerel's discovery encouraged many scientists to learn more about this radiation. Among these scientists were the Curies. They decided to study pitchblende to learn what was giving off radiation. They knew uranium was one source of the radiation, but the amount they found was too great to come from uranium only.

The Curies purchased pitchblende by the ton. They slowly purified the ore, getting rid of sand, clay, and other elements in the ore. After months of work, they finally isolated an element that had never been seen before. Marie Curie suggested the name polonium, in honor of her homeland, Poland. Polonium is hundreds of times more radioactive than uranium.

Physical properties

Polonium metal has a melting point of 254°C (489°F), a boiling point of 962°C (1,764°F), and a density of 9.4 grams per cubic centimeter.

Chemical properties

Polonium has chemical properties like the elements above it in the periodic table, especially selenium and tellurium. Polonium's chemical properties are of interest primarily to research chemists. Under most circumstances, scientists are more interested in polonium as a radioactive material.

Occurrence in nature

Polonium is produced in nature when other radioactive elements break down. It is so rare, however, that all the polonium needed is now made in particle accelerators.

Isotopes

Polonium has 27 isotopes, more than any other element. All of these isotopes are radioactive. Isotopes are two or more forms of an element. Isotopes differ from each other according to

WORDS TO KNOW

Alpha particles tiny, atom-sized particles that can destroy cells

Isotopes two or more forms of an element that differ from each other according to their mass number

Particle accelerator ("atom smasher") a machine used to cause small particles, such as protons, to move at very high speeds

Radioactivity having the tendency to break apart and give off some form of radiation

Toxic poisonous

their mass number. The number written to the right of the element's name is the mass number. The mass number represents the number of protons plus neutrons in the nucleus of an atom of the element. The number of protons determines the element, but the number of neutrons in the atom of any one element can vary. Each variation is an isotope.

A radioactive isotope is one that breaks apart and gives off some form of radiation. About the only isotope of any use is polonium-210.

Extraction
Polonium occurs so rarely and has so few uses that it is extracted from natural ores only for the purpose of research.

Uses
Polonium releases a great deal of energy during its radioactive breakdown. This has led to the development of compact heat sources for specialized purposes, such as use on space probes.

*The energy released by polonium during its radioactive breakdown is used in compact heat sources in space probes. This is the **Mariner 10**, launched November 3, 1973, on the first trip to the planet Mercury.*

Radiation is used to remove static electricity from photographic film. Static electricity consists of electric charges that collect on the outside of a surface. In film, they can reduce the clarity of a picture. The radiation polonium releases creates electrical charges in the air around it. These charges combine with those on the film, neutralizing them and preventing damage to the film.

Compounds
There are no compounds of polonium of practical interest. Some polonium compounds are prepared for the purpose of research.

Health effects
Polonium is an extremely dangerous substance. When it breaks down, it gives off alpha particles. These particles are tiny, atom-sized particles that can destroy cells. Polonium is considered to be more than 100 billion times more dangerous than hydrogen cyanide. The maximum suggested exposure to the element is no more than about seven one-hundred-billionths of a gram.

A relatively new hazard of polonium has recently been identified. The element has been found in the tobacco used in cigarettes and other products. The amount of polonium taken in by a smoker is approximately equal to that taken in from all other sources. Polonium must be added, therefore, to the list of harmful chemicals inhaled during smoking.

Polonium is an extremely dangerous material. It has recently been found in the tobacco used in cigarettes.

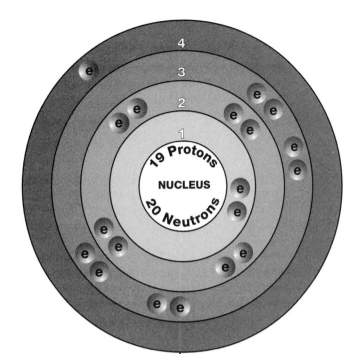

POTASSIUM

Overview

Potassium is one of the alkali metals. The alkali metals are the elements that make up Group 1 (IA) of the periodic table. The periodic table is a chart that shows how chemical elements are related to one another. The alkali metals also include **lithium, sodium, rubidium, cesium,** and **francium.** They are among the most active metals.

Potassium is so active that it never occurs free in nature. It always occurs in compounds, combined with other elements. It was first prepared in pure form in 1807 by English chemist Sir Humphry Davy (1778–1829). Davy used a new method of isolating elements that he had invented, electrolysis. In electrolysis, an electric current is passed through a molten (melted) compound. The electrical current breaks the compound into its elements. (See sidebar on Davy in the **calcium** entry in Volume 1.)

There are very few uses for potassium as a pure element. However, compounds of potassium have many important applications, the most important of which is as a fertilizer.

Discovery and naming

Early humans were familiar with potash, a potassium compound that forms when wood burns. Wood ashes were washed

SYMBOL
K

ATOMIC NUMBER
19

ATOMIC MASS
39.0983

FAMILY
Group 1 (IA)
Alkali metal

PRONUNCIATION
poe-TAS-see-um

with water to dissolve the potash. It was then recovered by evaporating the water. Potash was often called vegetable alkali. That name comes from the origin of the material ("vegetable" plants that contain wood) and the most important property of the material, alkali. The word alkali means a strong, harsh chemical that can be used for cleaning. Common household lye (such as Drano) is a typical alkali.

The chemical name for potash is potassium carbonate (K_2CO_3). Early humans also knew about a similar substance called mineral alkali. This material was made from certain kinds of rocks. But it also had alkali properties. "Mineral alkali" was also called soda ash. The modern chemical name for soda ash is sodium carbonate (Na_2CO_3).

For many centuries, people had trouble telling "vegetable alkali" and "mineral alkali" apart. The two materials looked and acted very much alike. For example, they could both be used as cleaning materials. The main difference between them was the source from which they came. It was not until the eighteenth century that chemists understood the difference between potash (vegetable alkali) and soda ash (mineral alkali).

By the late 1700s, chemists were reasonably sure that both potash and soda ash contained elements they had never seen. They tried to think of ways to break these compounds down into their elements. The first method that Davy tried was to pass an electric current through a water solution of one compound or the other. But no new element was ever formed. What Davy did not know was how active the elements potassium and sodium are. Both elements are freed when an electric current is passed through a water solution of potash or soda ash. But as soon as the element is formed, it reacts immediately with the water. The free element can never be recovered by this method.

Then Davy thought of another way to separate potash and soda ash into their elements. He decided to use no water in his experiment. Instead, he melted a sample of potash and a sample of soda ash. Then he passed an electric current through the molten (melted) substances. He was amazed to see a tiny liquid droplet of metal formed in each case. The droplet was the first piece of potassium and sodium ever to be seen by a human.

WORDS TO KNOW

Alkali metal an element in Group 1 (IA) of the periodic table

Isotopes two or more forms of an element that differ from each other according to their mass number

Periodic table a chart that shows how the chemical elements are related to each other

Potash a potassium compound that forms when wood burns

Radioactive isotope an isotope that breaks apart and gives off some form of radiation

Davy had his first success with potassium using this approach on October 6, 1807. A few days later he repeated his experiment with soda ash and produced pure sodium metal. Davy named these two elements after their much older names: potassium for "potash" and sodium for "soda ash."

Physical properties

Potassium is a soft, silvery-white metal with a melting point of 63°C (145°F) and a boiling point of 770°C (1,420°F). Its density is 0.862 grams per cubic centimeter, less than that of water (1.00 grams per cubic centimeter). That means that potassium metal can float on water. Chemically, though, that's not a good idea (*see* "Chemical properties" below).

The melting point of potassium is very low for a metal. It will melt over the flame of a candle flame.

Chemical properties

Like the other alkali metals, potassium is very active. It reacts with water violently and gives off **hydrogen** gas:

$$2K + 2H_2O \rightarrow 2KOH + H_2$$

So much heat is produced in this reaction that the hydrogen gas actually catches fire and may explode. Floating potassium metal on the surface of water is not a good idea! In that instance, the potassium would skip along the surface of the water. The skipping is caused by hydrogen gas produced in the reaction pushing the metal around. The potassium would soon catch fire, burn, and, perhaps, explode.

Potassium reacts readily with all acids and with all non-metals, such as **sulfur, chlorine, fluorine, phosphorus,** and **nitrogen.**

Occurrence in nature

Potassium is the eighth most abundant element in the Earth's crust. Its abundance is estimated to be about 2.0 to 2.5 percent. It is just slightly less abundant than its alkali cousin, sodium.

Potassium occurs widely in many different minerals. Some of the most important of these are sylvite, or potassium chloride (KCl); sylvinite, or sodium potassium chloride (NaCl • KCl); carnallite, or potassium magnesium chloride (KCl • $MgCl_2$); lang-

Potassium is less dense than water, so it can float on water. However, chemically, potassium reacts with water violently. It will give off hydrogen and eventually catch fire.

 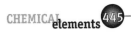

beinite, or potassium magnesium sulfate ($K_2SO_4 \bullet 2MgSO_4$); and polyhalite, or calcium magnesium potassium sulfate ($2CaSO_4 \bullet MgSO_4 \bullet K_2SO_4$).

Isotopes

There are three naturally occurring isotopes of potassium, potassium-39, potassium-40, and potassium-41. Potassium-40 is radioactive. Isotopes are two or more forms of an element. Isotopes differ from each other according to their mass number. The number written to the right of the element's name is the mass number. The mass number represents the number of protons plus neutrons in the nucleus of an atom of the element. The number of protons determines the element, but the number of neutrons in the atom of any one element can vary. Each variation is an isotope.

About ten artificial radioactive isotopes of potassium are known also. A radioactive isotope is one that breaks apart and gives off some form of radiation. Artificially radioactive isotopes are produced when very small particles are fired at atoms. These particles stick in the atoms and make them radioactive.

Potassium-40 is of special interest to scientists. Potassium is widely distributed in nature in plants, animals, and rocks. That means that nearly everything on Earth contains at least a tiny amount of radioactive potassium-40. That includes the human body! About 0.012 percent of the potassium in the human body is radioactive potassium-40. However, that is not enough radiation to cause any harm.

Radioactive potassium-40 in rocks can be used to measure the age of objects. When the isotope gives off radiation, it breaks down to an isotope of **argon:**

$$\text{potassium-40} \xrightarrow{\text{loses radiation}} \text{argon-40}$$

A scientist can analyze a rock to see how much potassium-40 and how much argon-40 it contains. The older the rock, the more argon-40 and the less potassium-40 it contains. The younger the rock, the more potassium-40 and the less argon-40 it contains.

Chemistry and the environment

Environmentalists often say that everything in nature is related. Here is a good example of that principle:

Potash was a widely used material in Colonial America. People used the compound to make soap, glass, and dozens of other products. At the time, potash was easy to get. All one had to do was burn a tree and collect potash from its ashes.

The only problem was that a single tree does not produce much potash. To get all the potash a family might need, one might have to burn dozens or hundreds of trees. Colonists did not worry too much about this problem. America in the 1700s was covered with trees. Few people thought about or cared about "saving the environ-

ment." If they ran out of trees, they just moved farther west.

One can imagine what America would have looked like if Colonists continued this practice. Fortunately, they did not have to. In the 1780s, French chemist Nicolas Le Blanc (1742–1806) invented an inexpensive method for making soda ash. Le Blanc's method used salt, or sodium chloride (NaCl); limestone, or calcium carbonate ($CaCO_3$); and coal (pure carbon). These three materials are all common and inexpensive. The Le Blanc method of making soda ash is quick, easy, and cheap. Before long, soda ash had become one of the least expensive chemicals made artificially. In the United States, trees were no longer burned to get potash.

One might wonder why argon gas does not escape into the atmosphere. The answer is that argon gas is trapped within the solid rock. It is released only when the potassium-dating process is conducted.

Extraction

The word "potash" is still a widely used term for potassium compounds taken from the earth. But it no longer means potassium carbonate to most people. It can mean potassium sulfate (K_2SO_4), potassium chloride (KCl), potassium nitrate (KNO_3), potassium hydroxide (KOH), or potassium oxide (K_2O). People cling to "potash" because it is the term used in the manufacture of fertilizers. And fertilizers are far and away the most important use of potassium compounds today.

The most important source of potash in the United States is a mine near Carlsbad, New Mexico, that produces sylvinite (KCl • NaCl). That mine produces about 85 percent of the potash mined in the United States. Potash is also produced from huge long-buried salt mine blocks formed when ancient seas evaporated (dried up). In Michigan, for example, potash is obtained

Potassium bicarbonate is used in the production of soft drinks.

by passing water into these mines. The water returns to the surface and is allowed to evaporate. Potassium compounds in the mine remain behind when the water has all evaporated.

Potassium metal is produced by combining potassium chloride with sodium metal at high temperatures. But this method is of little interest because potassium metal has few uses.

Uses

Potassium metal is sometimes used as a heat exchange medium. A heat exchange medium is a material that picks up heat in one place and carries it to another place. Potassium metal is sometimes used as a heat exchange medium in nuclear power plants. There, heat is produced at the core, or center, of the reactor. Liquid potassium is sealed into pipes surrounding the core. As heat is given off, it is absorbed (taken up) by the potassium. The potassium is then forced through the pipes into a nearby room. In that room, the potassium pipes are wrapped around pipes filled with water. The heat in the potassium warms the water.

Eventually the water gets hot enough to boil. It changes into steam and is used to operate devices that generate electricity.

Compounds

By far the most important compound of potassium is potassium chloride (usually referred to as "potash," of course!). At least 85 percent of that compound is used to make synthetic (artificial) fertilizers. In 1996, about 1.5 billion kilograms (3.4 billion pounds or 1.7 million tons) of potassium chloride was produced in the United States for use in fertilizers.

Many other potassium compounds are commercially important, although no use begins to compare with the amount of potash used for fertilizers. Some examples of other important potassium compounds are the following:

potassium bicarbonate, or baking soda ($KHCO_3$): baking powders; antacid (for upset stomach); food additive; soft drinks; fire extinguishers

potassium bisulfite ($KHSO_3$): food preservative (but not in meats); bleaching of textiles and straw; wine- and beer-making; tanning of leathers

potassium bitartrate, or cream of tartar ($KHC_4H_4O_6$): baking powder; "tinning" of metals; food additive

potassium bromide (KBr): photographic film; engraving

potassium carbonate, or potash (K_2CO_3): specialized glasses and soaps; food additive

potassium chromate (K_2CrO_4): dyes and stains (bright yellowish-red color); explosives and fireworks; safety matches; tanning of leather; fly paper

potassium fluorosilicate (K_2SiF_6): specialized glasses, ceramics, and enamels; insecticide

potassium hydroxide, or caustic potash (KOH): paint remover; manufacture of specialized soaps; fuel cells and batteries; bleaching; food additive; herbicide

potassium nitrate, or nitre, or saltpeter (KNO_3): explosives, fireworks, matches, rocket fuel; manufacture of glass; curing of foods

Potassium is one of the three primary nutrients, or macronutrients, required by plants.

Potassium bicarbonate is an additive in fire extinguishers.

potassium pyrophosphate, or tetrapotassium pyrophosphate, or TKPP ($K_4P_2O_7$): soaps and detergents

potassium sodium tartrate, or Rochelle salt ($KNaC_4H_4O_6$): baking powder; medicine; silvering of mirrors

Health effects

Potassium is essential to both plant and animal life. It is one of the three primary nutrients, or macronutrients, required by plants. Plants require relatively large amounts of potassium in order to grow and remain healthy.

Potassium plays a number of important roles in the human body also. It helps control the proper balance of fluids in cells and body fluids. It is involved in the transmission of chemical messages between nerve cells and in the contraction of muscles. Potassium also helps in the digestion of food and in the proper function of the eyes. In many of these reactions, potassium and sodium work together to keep these functions performing properly.

The average human who weighs 70 kilograms (150 pounds) has 140 grams (5 ounces) of potassium in his or her body. Normal daily intake of potassium is about 3.3 grams (0.1 ounce). Since potassium occurs in all plants, humans normally do not have any problems getting enough of the element in their daily diet.

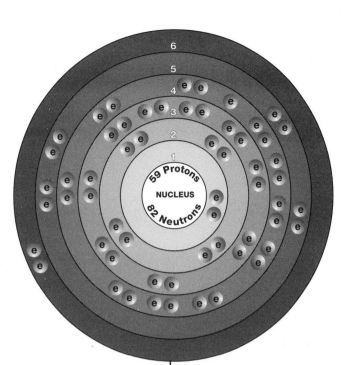

PRASEODYMIUM

Overview

During the late 1830s and early 1840s, Swedish chemist Carl Gustav Mosander (1797–1858) was studying two puzzling minerals, ceria and yttria. Both minerals had been discovered more than fifty years earlier in remote parts of Sweden. The minerals were puzzling because they seemed to consist of a mixture of new elements. Mosander eventually showed that one of the elements in ceria produced pink compounds. He called the new element didymium.

A few years after didymium was discovered, Austrian chemist Carl Auer (Baron von Welsbach) (1858–1929) made a correction to Mosander's research. Didymium was not a pure element, Auer announced, but a combination of two other new elements. He called these elements **neodymium** and praseodymium.

Praseodymium lies in Row 6 of the periodic table. The periodic table is a chart that shows how chemical elements are related to each other. The elements that make up Row 6 are sometimes called the rare earth metals. But the term is not very accurate. The rare earth elements are not especially rare in the Earth's crust. They were given this name because they have very similar properties. This similarity makes them difficult to

SYMBOL
Pr

ATOMIC NUMBER
59

ATOMIC MASS
140.9077

FAMILY
Lanthanide
(rare earth metal)

PRONUNCIATION
PRAY-zee-oh-DIM-ee-um

separate from each other. A better name for the rare earth elements is the lanthanides. This name comes from the first element in Row 6, **lanthanum.**

Praseodymium is a typical metal, somewhat similar to **aluminum, iron,** or **magnesium.** It is quite expensive to prepare and does not have many practical uses.

Discovery and naming

Mosander had been educated as a physician and a pharmacist. In the early 1830s, he was put in charge of the minerals collection at the Stockholm Academy of Sciences. He became very interested in two minerals that had been discovered in Sweden many years before, yttria and cerite. He devoted many years to studying the composition of these two minerals.

In 1841, Mosander announced that he had obtained two new elements from cerite. He called these elements lanthanum and didymium. He was correct about lanthanum being a new element, but he was wrong about didymium. This new "element" turned out to be a mixture of two other new elements, now called neodymium and praseodymium.

The man who made this discovery was Auer. He selected these two names because they mean "new twin" (neodymium) and "green twin" (praseodymium). The elements were called "twins" because they were both so much like lanthanum.

The praseodymium prepared by Auer was not very pure. It was contaminated with other elements. The first really pure sample of praseodymium was not made until 1931.

Physical properties

Praseodymium is a soft, malleable, ductile metal with a yellowish, metallic shine. Malleable means capable of being hammered into a thin sheet. Ductile means capable of being made into thin wires. Praseodymium has a melting point of 930°C (1,710°F) and a boiling point of about 3,200°C (about 5,800°F). Its density is 6.78 to 6.81 grams per cubic centimeter. Two allotropes of praseodymium exist. Allotropes are forms of an element with different physical and chemical properties. One allotrope, the "alpha" form, changes into a second allotrope, the "beta" form, at about 800°C.

WORDS TO KNOW

Allotropes forms of an element with different physical and chemical properties

Ductile capable of being drawn into thin wires

Isotopes two or more forms of an element that differ from each other according to their mass number

Lanthanides the elements in the periodic table with atomic numbers 58 through 71

Malleable capable of being hammered into thin sheets

Periodic table a chart that shows how chemical elements are related to each other

Pyrophoric gives off sparks when scratched

Rare earth elements *see* **Lanthanides**

Chemical properties

When it becomes moist, praseodymium reacts with **oxygen** in air to form praseodymium oxide. Praseodymium oxide (Pr_2O_3) forms as a greenish-yellow scale (like rust) on the surface of the metal. To protect praseodymium for this reaction, it is stored under mineral oil or covered with a plastic wrap.

Like many other metals, praseodymium also reacts with water and with acids. In these reactions, **hydrogen** gas is released.

Occurrence in nature

Praseodymium is one of the more common lanthanides. It is thought to occur with an abundance of about 3.5 to 5.5 parts per million in the Earth's crust. It occurs primarily with the other rare earth elements in two minerals, monazite and bastnasite.

Isotopes

Only one naturally occurring isotope of praseodymium is known. Isotopes are two or more forms of an element. Isotopes differ from each other according to their mass number. The number written to the right of the element's name is the mass number. The mass number represents the number of protons plus neutrons in the nucleus of an atom of the element. The number of protons determines the element, but the number of neutrons in the atom of any one element can vary. Each variation is an isotope.

At least 15 radioactive isotopes of praseodymium are known also. A radioactive isotope is one that breaks apart and gives off some form of radiation. Radioactive isotopes are produced when very small particles are fired at atoms. These particles stick in the atoms and make them radioactive. None of the radioactive isotopes has any commercial use.

Extraction

The first step in obtaining praseodymium is to treat monazite, bastnasite, or another ore to separate the lanthanides from each other. The various elements are then changed to compounds of **fluorine,** such as praseodymium fluoride (PrF_3). Praseodymium metal can then be obtained by passing an electric current through praseodymium fluoride:

$$2PrF_3 \xrightarrow{\text{electric current}} 2Pr + 3F_2$$

The original "discovery" of a "new" element called didymium turned out to be a mixture of neodymium and praseodymium.

Praseodymium

A woman during World War II wears welders' goggles. Praseodymium is a component of the glass used in these safety goggles.

or by making it react with an active metal:

$$3Na + PrF_3 \rightarrow 3NaF + Pr$$

It is still quite expensive to make praseodymium, and the metal sells for about $1,200 a kilogram ($2,500 a pound).

Uses

One of the oldest uses for praseodymium is in the manufacture of misch metal. Misch metal is pyrophoric, meaning that the metal gives off sparks when it is scratched. The most common use of misch metal is in lighter flints and tracer bullets. When a metal wheel is rubbed across misch metal in a cigarette lighter, the metal gives off sparks. Those sparks then set fire to lighter fluid, giving a flame to light a cigarette.

Like other lanthanides, praseodymium is also used to give color to glass, ceramics, enamels, and other materials. The characteristic color provided by compounds of praseodymium is a bright yellow.

A related use of praseodymium is in carbon arc lamps, like those used in the motion picture industry. When an electric current is passed through a carbon arc, the arc gives off a brilliant white light. The addition of a small amount of praseodymium gives a brilliant yellow cast to the light.

Praseodymium is also a component of didymium glass. Didymium glass contains a mixture of rare earth elements, including lanthanum, praseodymium, neodymium, **samarium, cerium,** and **gadolinium.** This glass is used to make welder's goggles. It helps protect the welder's eyes from the intense light produced during welding.

Compounds
Relatively few compounds of praseodymium have any commercial uses.

Health effects
The health effects of praseodymium are not well studied. As a safety measure, chemists treat the metal as if it were toxic and handle it with caution.

Praseodymium is used to give a bright yellow color to glass, ceramics, and enamels.

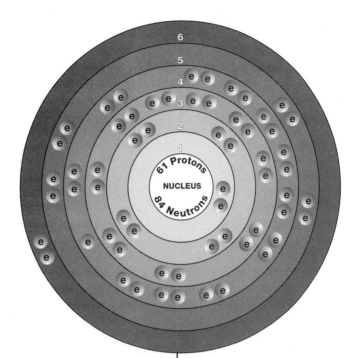

PROMETHIUM

Overview

Promethium is one of the most fascinating of all chemical elements. It has never been found on the Earth's surface. Scientists know of it only because it can be prepared artificially in particle accelerators ("atom smashers") and in other unusual reactions. Its existence was predicted as early as 1902, but its discovery was not confirmed until 1945.

All of the known isotopes of promethium are radioactive. That is, they break down and give off radiation spontaneously.

At one time, promethium was strictly a laboratory curiosity. Today, however, it has a number of practical industrial applications.

SYMBOL
Pm

ATOMIC NUMBER
61

ATOMIC MASS
144.9128

FAMILY
Lanthanide
(rare earth metal)

PRONUNCIATION
pruh-MEE-thee-um

Discovery and naming

In the late 1860s, Russian chemist Dmitri Mendeleev (1834–1907) discovered the periodic law. The periodic law provides a way of organizing the chemical elements to show how they are related to each other. It is usually represented by a table with 18 columns and 7 rows. Each chemical element belongs in one of the boxes of the periodic table.

By about 1900, most of the chemical elements had been discovered, but a few empty boxes remained on the periodic table. Chemists wondered why those boxes were still empty. In 1902, Czech chemist Bohuslav Brauner predicted that there should be an element between **neodymium** (number 60) and **samarium** (number 62). Chemists began searching for the element based on the characteristics of the elements around it.

In 1924, Italian chemists Luigi Rolla and Rita Brunetti claimed to have found element 61. They suggested the name florentium for their home town of Florence. At about the same time, scientists at the University of Illinois also announced the discovery of element 61. They proposed the name illinium for Illinois.

Gradually, scientists began to believe that element 61 was radioactive. A radioactive element is one that breaks apart and gives off some form of radiation. One way to make radioactive elements is to fire very small particles at atoms. The particles stick in the atoms and make them radioactive. In the late 1930s, scientists at Ohio State University thought they had found element 61. They suggested the name cyclonium, after the kind of particle accelerator they used to make the element, a cyclotron.

None of the "discoveries"—from Italy, Illinois, or Ohio—could be confirmed by other scientists. A great debate went on for many years as to whether element 61 had really been found or not. Finally, the problem was solved. During World War II (1939–45), scientists at the Oak Ridge Laboratory in Oak Ridge, Tennessee, were studying the materials formed during atomic fission. Atomic fission is the process in which large atoms break apart, releasing large amounts of energy and smaller atoms. The smaller atoms are called fission products.

The Oak Ridge scientists proved that element 61 was present in fission products of **uranium.** They named it promethium, after the Greek god Prometheus. According to legend, Prometheus stole fire from the gods and brought it to Earth for human use.

Physical properties

Promethium is a silver-white metal with a melting point of 1,160°C (2,120°F) and no measured boiling point. Its density is 7.2 grams per cubic centimeter. The physical properties of promethium are of less interest to scientists than its radioactive properties.

WORDS TO KNOW

Atomic fission the process in which large atoms break apart producing large amounts of energy and smaller atoms

Isotopes two or more forms of an element that differ from each other according to their mass number

Luminescence the property of giving off light without giving off heat

Periodic law a way of organizing the chemical elements to show how they are related to each other

Radioactivity having the tendency to break apart and give off some form of radiation

Spectrum (plural spectra): the pattern of light given off by a glowing object, such as a star

Chemical properties

Promethium behaves like other rare earth elements. The chemical properties of promethium are of less interest to scientists than its radioactive properties.

Occurrence in nature

Promethium has never been found in the Earth's crust. It has been observed, however, in the spectra of some stars in the galaxy of Andromeda. The spectrum (plural: spectra) of a star is the light given off by the star.

The Andromeda galaxy. Promethium has been observed in the spectra of some stars in this galaxy.

Isotopes

Fifteen isotopes of promethium are known. Isotopes are two or more forms of an element. Isotopes differ from each other according to their mass number. The number written to the right of the element's name is the mass number. The mass number represents the number of protons plus neutrons in the nucleus of an atom of the element. The number of protons determines the element, but the number of neutrons in the atom of any one element can vary. Each variation is an isotope.

The only isotope generally available is promethium-147, with a half life of 2.64 years. The half life of a radioactive element is the time it takes for half of a sample of the element to break down. That means for promethium-147 that after 2.64 years, only half of a 100-gram sample, for example, or 50 grams, will be left. Another isotope, promethium-145, has a longer half life of 18 years.

Extraction

Promethium is not found in the Earth's surface.

Uses and compounds

Promethium has limited uses. It can be used as a source of power. The radiation it gives off provides energy, similar to that from a battery. A promethium battery can be used in places where other kinds of batteries would be too heavy or large to use, as on satellites or space probes. Such batteries are far too expensive for common use, however.

Promethium is also used to measure the thickness of materials. For example, suppose thin sheets of metal are being produced on a conveyor belt. A sample of promethium metal is placed above the metal and a detector is placed below. The detector counts the amount of radiation passing through the metal. If the metal sheet becomes too thick, less radiation passes through. If the sheet becomes too thin, more radiation passes through. The detector reports when the sheet of metal is too thick or too thin. It can automatically stop the conveyor belt when this happens.

Some compounds of promethium are luminescent. Luminescence is the property of giving off light without giving off heat. The light of a firefly is an example of luminescence.

Promethium compounds are luminescent because of the radiation they give off.

Health effects
Like all radioactive materials, promethium must be handled with great care. The radiation it produces can have serious health effects on humans and animals.

PROTACTINIUM

Overview

Protactinium is one of the rarest elements on Earth. It is formed when **uranium** and other radioactive elements break down. For many years, the only supply of protactinium of any size was kept in Great Britain. The British government had spent $500,000 to extract 125 grams (about four ounces) of the element from 60 tonnes (60 tons) of radioactive waste. Relatively little is known about the properties of the element, and it has no commercial uses.

Protactinium belongs in the actinides series in the periodic table. The periodic table is a chart that shows how chemical elements are related to one another.

Discovery and naming

Scientists first learned about radioactive elements toward the end of the nineteenth century. Radioactive elements are elements that break apart all by themselves. They give off radiation—somewhat similar to light or X rays—and change into new elements. Radiation is energy transmitted in the form of electromagnetic waves or subatomic particles.

For example, the element uranium is radioactive. It emits radi-

SYMBOL
Pa

ATOMIC NUMBER
91

ATOMIC MASS
231.03588

FAMILY
Actinide

PRONUNCIATION
pro-tack-TIN-ee-um

ation over very long periods of time. It begins to change into other elements. One of those elements is protactinium.

Many naturally occurring isotopes are radioactive. Isotopes are two or more forms of an element. Isotopes differ from each other according to their mass number. The number written to the right of the element's name is the mass number. The mass number represents the number of protons plus neutrons in the nucleus of an atom of the element. The number of protons determines the element, but the number of neutrons in the atom of any one element can vary. Each variation is an isotope.

Many of the radioactive isotopes that occur in nature are related to each other. For example, when uranium-238 breaks apart, it forms a new isotope, **thorium**-238. But thorium-234 is radioactive also. It breaks apart to form **radium**-230. And radium-230 is also radioactive. It breaks apart to form **actinium**-230.

This process often goes on for a dozen steps or more. Finally, an isotope is formed that is not radioactive. The chain—or "family" of radioactive isotopes—comes to an end.

During the early 1900s, scientists were trying to understand these radioactive families. They were trying to identify all the elements found in a family. In doing so, they sometimes found new elements. Such was the case with element number 91. Many scientists had been looking for element number 91 for some time. There was an empty box in the periodic table for element 91. That meant that a new element was yet to be found. Some scientists decided to look in the radioactive families for that element.

In 1913 German-American physicist Kasimir Fajans (1887–1975) and his colleague, O. H. Göhring, claimed to have found element number 91. They suggested the name brevium for the element. They chose the name because the half life of the isotope they found was very short ("brief"). It was only 1.175 minutes.

The half life of a radioactive element is the time it takes for half of a sample of the element to break down. That means that 10 grams of the isotope they studied would break down very quickly. Only 5 grams would be left after 1.175 minutes. Then 2.5 grams (half of 5 grams) would be left after another

WORDS TO KNOW

Half life the time it takes for half of a sample of a radioactive element to break down

Isotopes two or more forms of an element that differ from each other according to their mass number

Periodic table a chart that shows how chemical elements are related to each other

Radiation energy transmitted in the form of electromagnetic waves or subatomic particles

Radioactive isotope an isotope that breaks apart and gives off some form of radiation

Radioactive family a group of radioactive elements and isotopes that are related to each other

German chemist Lise Meitner. She discovered an isotope of protactinium.

1.175 minutes, and 1.25 grams (half of 2.5 grams) after another 1.175 minutes, and so on. Until the discovery by Fajans and Göhring, the element had been known as uranium-X_2. That name came from the element's position in one of the radioactive families.

In 1918, another isotope of element number 91 was discovered by German physicists Lise Meitner (1878–1968) and Otto Hahn (1879–1968). This isotope had a half life of 32,500 years. It was much easier to study than the isotope discovered by Fajans and Göhring.

The new element was originally given the name protoactinium, meaning "first actinium." It comes from the way the element breaks down. Its first product is the element actinium. In 1949, the element's name was changed slightly to its current form, protactinium.

Physical properties

Protactinium is a bright shiny metal. When exposed to air, it combines easily with **oxygen** to form a whitish coating of pro-

German chemist Otto Hahn was involved in early work with protactinium.

tactinium oxide. Its melting point is thought to be about 1,560°C (2,840°F) and its density about 15.37 grams per cubic centimeter.

Chemical properities

Protactinium forms compounds with the halogens (**fluorine, chlorine, bromine,** and **iodine**) and with **hydrogen.** But these compounds have not been studied in detail.

Occurrence in nature

The amount of protactinium in the Earth's crust is too small to estimate accurately. Its most common ore, pitchblende, contains about 0.1 part per million of protactinium.

Isotopes

About 20 isotopes of protactinium are known. All are radioactive. (*See* "Discovery and naming" for a more detailed explanation of isotopes.)

Extraction

Protactinium does not occur naturally.

Lise Meitner | Austrian physicist

Until recently, science has often been a difficult occupation for women. Male scientists once believed that women did not have the mental powers to do good research. Women who became famous scientists usually had to be outstanding in their own field, and they had to overcome the strange prejudices of their male colleagues.

No one knew more about discrimination in science than Lise Meitner (1878–1968). Meitner was born in Vienna, Austria, on November 7, 1878. She learned about the work of Marie Curie while in high school and decided to pursue a career in science. She earned her Ph.D. degree in physics in 1906.

After working as a nurse during World War I (1914–18), Meitner took a job at the University of Berlin. At first, she had to overcome huge obstacles. Her superior would not allow her to work in a laboratory if men were present. He had a tiny laboratory built for her in a closet.

Meitner persevered, however. She eventually became a professor of physics at the school and also served as co-director of the Kaiser Wilhelm Institute in Berlin. The other co-director at this famous research institution was Otto Hahn (1879–1968), a physicist with whom Meitner worked throughout most of her career.

Meitner's career took an unexpected turn in the 1930s. When Adolf Hitler (1889–1945) came to power in Germany, he began to rid the universities of anyone with a Jewish background. Although Meitner had been baptized as a Christian, she came from a Jewish family. She soon realized that her life would be in danger if she stayed in Berlin. So she escaped from Germany in 1938 and took a position in Copenhagen, Denmark.

One discovery for which Meitner and Hahn are known is the discovery of protactinium. They found the element while searching through the products of a nuclear reaction that had only recently been discovered. In fact, the ability of Hahn and Meitner to unravel the nature of that reaction proved to be even more important than the discovery of protactinium.

The reaction in question was one that occurs when neutrons (tiny particles that occur in atoms) are fired at uranium atoms. The reaction had been carried out by a number of scientists, but only Meitner and Hahn figured out what had actually taken place. In 1939, they wrote a paper explaining the reaction. They said that neutrons caused uranium atoms to fission, or split apart.

Meitner and Hahn had described for the first time one of the most important reactions in all of human history: nuclear fission. Nuclear fission later became the basis for weapons, such as the atomic bomb, and useful applications, such as nuclear power plants. For his role in this discovery, Hahn was awarded a share of the 1944 Nobel Prize in Chemistry. Meitner, who had contributed at least as much as Hahn, never received a Nobel Prize for her work. Scholars are still debating the reasons that Meitner's brilliant work was ignored by the Nobel Prize committee in 1944.

Uses and compounds

Neither protactinium nor its compounds have any commercial uses. It can be purchased in small amounts today from the Oak

Ridge National Laboratory in Oak Ridge, Tennessee. It costs about $300 per gram.

Health effects

Protactinium is very radioactive and highly dangerous. Researchers who work with it must take extreme cautions to protect themselves from its radiation.

RADIUM

Overview

Radium is a radioactive element in Group 2 (IIA) and Row 7 of the periodic table. The periodic table is a chart that shows how chemical elements are related to each other. Radium was discovered in 1898 by Marie Curie (1867–1934) and her husband, Pierre Curie (1867–1934). It was found in an ore of **uranium** called pitchblende. The alkaline earth metals also include **beryllium, magnesium, calcium, strontium,** and **barium.**

Radium is luminescent, meaning it gives off radiation that can be seen in the dark. Because of its radiation, however, it has relatively few uses.

Discovery and naming

The discovery of radium is one of the most interesting stories in science. The story has been told over and over again in books, articles, and motion pictures, and on television.

The story begins with the research of French physicist Antoine-Henri Becquerel (1852–1908). In 1896, Becquerel made a discovery about the ore called pitchblende. Pitchblende contains the element uranium. Becquerel found that pitchblende gives off radiation that acts much like light. The

SYMBOL
Ra

ATOMIC NUMBER
88

ATOMIC MASS
226.0254

FAMILY
Group 2 (IIA)
Alkaline earth metal

PRONUNCIATION
RAY-dee-um

Polish-French physicist Marie Curie.

main difference is that the radiation from pitchblende is not visible to the human eye.

Becquerel's discovery caused great excitement among scientists. Many physicists stopped their own research and began to study this new curiosity. One of those who did so was a graduate student named Marie Sklodowska Curie. Marie had been born in Warsaw, Poland, as Marya Sklodowska. In 1891, she moved to Paris, France, to study physics. Three years later she met another physicist, Pierre Curie. The two were married in 1895.

What time is it?

Radium was once used in paint that was applied on the hands and numbers of clocks and watches. The visible radiation it emitted made it possible to read the numbers in the dark. But the radiation proved very harmful to people who applied the radium paint to a watch or clock.

The technique they used was to make a sharp point on their brushes by twirling the brush between their lips. They then dipped the brush into radioactive radium paint. The dipping and twirling sequence ultimately caused the painters to get a lot of radium on their lips. This resulted in many cases of lip and mouth cancer among those painters. So radium is no longer used on clocks and watches.

Marie and Pierre were especially interested in learning more about pitchblende. What was in the ore that was giving off radiation, they asked. To answer this question, they purified huge amounts of the natural ore. Eventually, they isolated a new element that gave off radiation much more intensely than did the pitchblende itself. The Curies named the new element **polonium.**

But they were not finished with their research. They thought at least one other element might be in the pitchblende. So they continued the process of purification. In 1898, they isolated a second new element. They called this element radium. They chose this name because the element gives off such intense radiation. It took the Curies another four years to prepare one gram of the element. To do so, they had to sift through more than seven metric tons of pitchblende!

Physical properties
Radium is a brilliant white metal with a melting point of 700°C (1,300°F) and a boiling point of 1,737°C (3,159°F). Its density is 5.5 grams per cubic centimeter.

Chemical properties
Radium combines with most non-metals, including **oxygen, fluorine, chlorine,** and **nitrogen.** It also reacts with acids with the formation of **hydrogen** gas. Radium's chemical properties are of much less interest than its radioactivity, however.

WORDS TO KNOW

Half life the time it takes for half of a sample of a radioactive element to break down

Isotopes two or more forms of an element that differ from each other according to their mass number

Periodic table a chart that shows how chemical elements are related to each other

Radioactive isotope an isotope that breaks apart and gives off some form of radiation

French physicist Pierre Curie.

Occurrence in nature

The amount of radium in the Earth's crust is very small. Its abundance has been estimated to be about 0.0000001 parts per million. It occurs not only in pitchblende, but in all ores that contain uranium. It is formed when uranium gives off radiation and breaks down.

Isotopes

Four naturally occurring isotopes of radium are known. They are radium-223, radium-224, radium-226, and radium-228. Isotopes are two or more forms of an element. Isotopes differ from each other according to their mass number. The number written to the right of the element's name is the mass number. The mass number represents the number of protons plus neutrons in the nucleus of an atom of the element. The number of protons determines the element, but the number of neutrons in the atom of any one element can vary. Each variation is an isotope.

Only radium-226 has any commercial applications. It has a half life of 1,620 years. After that period of time, only half of the

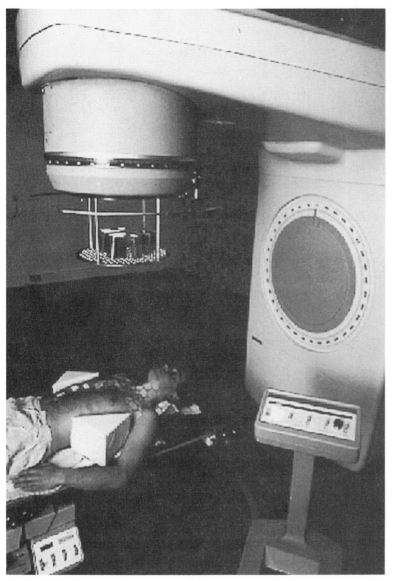

A man undergoes radiation treatment.

original sample would remain. The half life of a radioactive element is the time it takes for half of a sample of the element to break down. The other three isotopes have half lives of only a few days or years. These short half lives make it difficult to work with the isotopes.

The only isotope now used very often, radium-226, is generally not used directly. Instead, it is used to make radon gas. Radon gas is one of the products formed when radium breaks down. The radon gas is easier and safer to work with than is the radium itself.

Extraction

The process by which radium is obtained is similar to that used by the Curies. The metal is separated from other substances found in pitchblende by a long series of chemical reactions.

Uses and compounds

Because of its history, radium is a very interesting and important element. But radium and its compounds have relatively few uses. In fact, no more than about two kilograms (five pounds) of radium is made each year. The small amount of radium that is available is used for medical purposes. The radium is used to produce **radon** gas which, in turn, is used to treat cancer. Radiation given off by radium is sometimes used also to study the composition of metals, plastics, and other materials.

Health effects

Like all radioactive materials, radium is a dangerous substance to handle. The radiation it gives off can kill living cells. This property is desirable in treating cancer. Killing cancer cells can help a patient recover from the disease. But great care must be taken in using radium for this purpose. Its radiation can also kill healthy cells. People who work with radium must take great care that they do not get the element on their skin, swallow it, or inhale its fumes.

Marie Curie herself eventually died from working with radium. She developed leukemia and died in 1934.

6
5
4
3
2
1

86 Protons

NUCLEUS

136 Neutrons

RADON

Overview

Radon is the last member of the noble gas family. The noble gases are the elements that make up Group 18 (VIIIA) of the periodic table. The periodic table is a chart that shows how chemical elements are related to one another. The noble gases get their name because they are inactive chemically. They combine with other substances under only extreme conditions. Their tendency to avoid contact with other elements was seen by early chemists as "royal" or "noble" behavior. The noble gases are also called the inert gases.

Radon is a radioactive element. A radioactive element is one that gives off radiation and breaks down to form a different element. Radon is formed when heavier radioactive elements, like **uranium** and **thorium,** break down. In turn, radon breaks down to form lighter elements, such as **lead** and **bismuth.**

Radon is a well-know air pollutant today. It is formed in rocks and soil where uranium is present. As a gas, radon tends to drift upward out of the ground. If a house or building has been built above soil containing uranium, radon may collect in the structure. The U.S. Environmental Protection Agency (EPA)

SYMBOL
Rn

ATOMIC NUMBER
86

ATOMIC MASS
222.0176

FAMILY
Group 18 (VIIIA)
Noble gas

PRONUNCIATION
RAY-don

regards the presence of radon in homes and offices as a serious health problem.

Discovery and naming

Radioactivity was discovered in 1896 by French physicist Antoine-Henri Becquerel (1852–1908). Becquerel observed that a photographic plate was exposed even in the dark when placed next to an ore called pitchblende. The explanation for this phenomenon was offered two years later by a colleague of Becquerel's, Polish-French chemist Marie Curie (1867–1934). Curie said that something in the pitchblende was giving off radiation. The radiation was similar to light in some ways. But it was also different, since it could not be seen. Curie suggested the name of radioactivity for this behavior.

Over the next decade, many scientists worked to find out more about radioactive materials. Curie and her husband, Pierre Curie (1859–1906), isolated two new radioactive elements, **polonium** and **radium.** In 1900, German physicist Friedrich Ernst Dorn (1848–1916) found a third radioactive element.

Dorn found this element because of an observation made by Curie. When radium is exposed to air, the air becomes radioactive. The Curies did not study this phenomenon further. However, Dorn did. Eventually he discovered that radium produces a gas when it breaks apart. The radioactive gas escapes into the air. The radioactivity of air exposed to radium is caused by this gas.

At first, Dorn called this radioactive gas radium "emanation". The term emanation refers to something that has been given off. Radium emanation, then, means something given off by radium. Dorn also considered the name of niton for the gas. This name comes from the Latin word *nitens,* which means "shining." Eventually, however, scientists decided on the modern name of radon. The name is a reminder of the source from which the gas comes, *rad*ium.

The proper location of radon in the periodic table was determined by Scottish chemist Sir William Ramsay (1852–1916). Ramsay was also involved in the discovery of three other noble gases, **neon, krypton,** and **xenon.** In 1903, Ramsay was able to determine the atomic weight of radon. He showed that it belonged beneath xenon in Group 18 (VIIIA) of the periodic table.

WORDS TO KNOW

Half life the time it takes for half of a sample of a radioactive to break down

Inert incapable of reacting with other substances

Isotopes two or more forms of an element that differ from each other according to their mass number

Noble gas an element in Group 18 (VIIIA) of the periodic table

Periodic table a chart that shows how chemical elements are related to each other

Radioactive isotope an isotope that breaks apart and gives off some form of radiation

French physicist Antoine Henri Becquerel.

Credit for the discovery of radon is often given to other scientists as well. In 1899, Robert B. Owens announced the presence of a radioactive gas that he named thoron. In 1903, French chemist André Louis Debierne (1874–1949) made a similar discovery. He named the gas actinon. Certainly, some credit for the discovery of element 86 can be shared among all these men.

Physical properties

Radon is a colorless, odorless gas with a boiling point of –61.8°C (–79.2°F). Its density is 9.72 grams per liter, making it about seven times as dense as air. It is the densest gas known. Radon dissolves in water and becomes a clear, colorless liquid below its boiling point. At even lower temperature, liquid radon freezes. As a solid, its color changes from yellow to orangish-red as the temperature is lowered even further. It is a dramatic sight since it also glows because of the intense radiation being produced.

Chemical properties

Radon was long thought to be chemically inert. The term inert means incapable of reacting with other substances. In the

Radon: the secret visitor

A dangerous stranger may be hiding in your home. You won't be able to see, smell, or hear the stranger. But it has the ability to cause cancer. That dangerous stranger is radon gas.

Radon is produced naturally when uranium breaks down. Uranium is a radioactive element that occurs naturally in the Earth's crust. It is a fairly common element and could be in the ground below your own home.

When uranium breaks down, it produces many different elements, including radium, thorium, bismuth, and lead. None of these elements is a threat since they all remain in the ground. But uranium also forms radon when it breaks down. And radon is a gas. It can float upward, out of the earth, and into the basement of your home.

In some respects, radon is a serious health hazard. It gives off radiation that can kill cells. But radon does not have a very long half life. It breaks down and disappears fairly quickly.

The problem is that it breaks down into elements that are solid. These include polonium-214, polonium-218, and lead-214. These elements are more of a threat to your health. If you inhale them, they may stick to the lining of your lungs. While there, they give off radiation. The radiation can kill or damage cells. The final result of radon escaping into a building can be a variety of respiratory problems. Respiratory problems are those affecting the lungs and other parts of the system used for breathing. The most serious of these respiratory problems is lung cancer.

Scientists today think that radon may cause as many as 20,000 cases of lung cancer per year. If so, that would make radon the second leading cause of this disease, after smoking. The people most in danger from radon are those who also smoke. These people are threatened both by radon and by cigarette smoke.

The EPA has studied the problem of radon in homes and offices. The agency believes that up to 8 million homes may have levels of radon that are too high. About 20 percent of all homes the agency has studied have high radon levels.

Fortunately, it's easy to find out if radon is lurking in your home. Radon test kits can be purchased easily and at low cost. Anyone can learn how to use one in a few minutes. If radon is present, some simple steps can be taken to reduce the danger the element presents. For example, any cracks in the foundation of a house can be sealed. By doing so, radon gas will be prevented from seeping into the house. Also, some method for circulating air should always be available. A fan or an air conditioner, for example, will insure that fresh air is constantly brought into a house and "stale" air (containing radon gas) is removed.

early 1960s, however, a number of chemists found ways of making compounds of the noble gases. They did so by combining a noble gas with a very active element. The element generally used was **fluorine,** the most active chemical element. The result was the formation of noble gas compounds. The first radon compound to be produced was radon fluoride (RnF).

Occurrence in nature

The abundance of radon in air is too small to be estimated. Some radon is always present because it is formed during the breakdown of uranium and radium.

Isotopes

Three isotopes of radon occur in nature—radon-219, radon-220, and radon-222. Isotopes are two or more forms of an element. Isotopes differ from each other according to their mass number. The number written to the right of the element's name is the mass number. The mass number represents the number of protons plus neutrons in the nucleus of an atom of the element. The number of protons determines the element, but the number of neutrons in the atom of any one element can vary. Each variation is an isotope. At least 18 other radioactive isotopes of radon have been produced artificially.

All isotopes of radon have short half-lives and do not remain in the atmosphere very long. The half life of a radioactive element or isotope is the time it takes for half of a sample of the element or isotope to break down. The radon isotope with the longest half life is radon-222 at only 2.8 days. If 10 grams of radon-222 were prepared today, only 5 grams would remain 2.8 days from now. After another 2.8 days, only 2.5 grams would be left. Within a month, it would be difficult to detect any of the isotope.

Extraction

Radon is produced during the breakdown of radium. It is obtained commercially by the following method. A compound of radium is placed under water. Gases given off by the radium compound are collected in a glass tube. **Oxygen, nitrogen,** water vapor, carbon dioxide, and other gases are removed from the gas in the tube. The gas that remains is pure radon.

Uses

The uses for radon all depend on the radiation it gives off. That radiation cannot be seen, smelled, tasted, or detected by any other human sense. However, a number of instruments have been invented for detecting this radiation. For example, a Geiger counter is a device that makes a clicking sound or flashes a light when radiation passes through it.

One use of radon based on this principle is in leak detection. An isotope of radon is added to a flow of gas or liquid through

As a solid, radon changes its color from yellow to orangish-red as its temperature decreases. It is a dramatic sight since it also glows because of the intense radiation being produced.

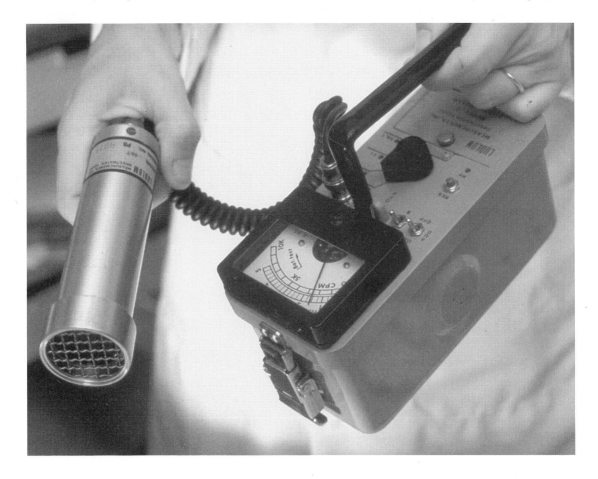

A handheld Geiger counter, used to detect radiation. Together with an isotope of radon, leaks can be located.

a tube. A Geiger counter can be passed along the outside of the tube. If radiation is present, the Geiger counter makes a sound or flashes a light. The presence of radiation indicates a leak in the tube. This principle is applied in many other systems to study materials that cannot actually be seen.

Radon was once commonly used to treat cancer too. The radiation it gives off kills cancer cells. However, the element must be used with great care because radiation can kill healthy cells as well. In fact, the bad side-effects of radiation therapy are caused by the killing of healthy cells by radiation. Today, radon is not as widely used for the treatment of cancer. More efficient isotopes have been found that are easier and safer to work with.

Compounds
Chemists are trying to make compounds of radon, but the task is difficult. One compound that *has* been made is radon fluoride. In any event, such compounds are laboratory curiosities and have no commercial uses.

Health effects

Because of the radiation it produces, radon is a highly dangerous material. It is used only with great caution. Radon is especially dangerous because it is inhaled, exposing fragile tissues to penetrating radiation.

RHENIUM

Overview

Rhenium was discovered by a German research team that included Walter Noddack (1893–1960), Ida Tacke (1896–1979) and Otto Berg. These scientists knew that there were two empty boxes in the periodic table that represented elements that had not yet been discovered. The periodic table is a chart that shows how chemical elements are related to one another. In 1925, the German team announced that they had found both elements. They were correct about one (element number 75) but wrong about the other (element number 43).

Rhenium is one of the rarest elements in the world. At one time it sold for about $10,000,000 a kilogram (about $5,000,000 a pound). It is no longer that expensive, although it is still very costly.

Rhenium has some unusual properties. For example, it is one of the most dense elements known. It also has one of the highest boiling points of all elements.

The primary uses of rhenium are in alloys that are used at very high temperatures or exposed to a great deal of wear.

SYMBOL
Re

ATOMIC NUMBER
75

ATOMIC MASS
186.207

FAMILY
Group 7 (VIIB)
Transition metal

PRONUNCIATION
REE-nee-um

Alloy a mixture of two or more metals with properties different from those of the individual metals

Catalyst a substance used to speed up or slow down a chemical reaction without undergoing any change itself

Ductile capable of being drawn into thin wires

Half life the time it takes for half of a sample of a radioactive element to break down

Isotopes two or more forms of an element that differ from each other according to their mass number

Malleable capable of being hammered into thin sheets

Periodic table a chart that shows how the chemical elements are related to each other

Radioactive having a tendency to break apart and give off some form of radiation

Superalloy an alloy made of iron, cobalt, or nickel that has special properties, such as the ability to withstand high temperatures and attack by oxygen

Toxic poisonous

Discovery and naming

At the beginning of the 1920s, chemists knew they were approaching a milestone. They had already isolated 87 chemical elements. But they knew that five more were waiting to be discovered. How did they know? Every element has a space in the periodic table. An empty space meant that an element was missing. In 1920, five empty spaces were still left in the periodic table.

Chemists worldwide were searching for these five elements. In 1925, Noddack, Tacke, and Berg reported that they had found two of those elements: numbers 43 and 75. They called the first element masurium, after the region called Masurenland in eastern Germany. They named element number 75 rhenium, after the Rhineland, in western Germany. Rhenium was the last naturally occurring element to be discovered.

When a discovery like this is announced, other chemists try to repeat the experiments. They see if they get the same results as those reported. In this case, the German team turned out to be half right. Scientists were able to confirm the existence of element 75. They were not able to confirm the Germans' discovery of element 43. In fact, it was another decade before element 43 (**technetium**) was actually discovered.

Physical properties

Rhenium is a ductile, malleable, silvery metal. Ductile means capable of being drawn into thin wires. Malleable means capable of being hammered into thin sheets. It has a density of 21.02 grams per cubic centimeter, a melting point of 3,180°C (5,760°F), and a boiling point of 5,630°C (10,170°F). These numbers are among the highest to be found for any of the chemical elements.

Rhenium is quite dense, which is unusual for a metal. When heated, most metals reach a point where they change from being ductile to being brittle. They can be worked with below that point, but not above it. Above this transition temperature they become brittle. If one tries to bend or shape them, they break apart. The unusual behavior of rhenium means that it can be heated and recycled many times without breaking apart.

Chemical properties

Rhenium is a moderately stable metal. It does not react with **oxygen** and some acids very readily. But it does react with strong acids such as nitric acid (HNO_3) and sulfuric acid (H_2SO_4).

Occurrence in nature

About a third of all rhenium used in the United States comes from **copper** and **molybdenum** ores in the Western states. It is obtained during the process of copper mining. Two-thirds are imported from other countries, primarily Chile, Germany, and the United Kingdom. The principal ores of rhenium are molybdenite, gadolinite, and columbite.

Rhenium is one of the rarest elements in the world. Its abundance is thought to be about one part per billion.

Isotopes

Two isotopes of rhenium occur in nature, rhenium-185 and rhenium-187. Isotopes are two or more forms of an element. Isotopes differ from each other according to their mass number. The number written to the right of the element's name is the mass number. The mass number represents the number of protons plus neutrons in the nucleus of an atom of the element. The number of protons determines the element, but the number of neutrons in the atom of any one element can vary. Each variation is an isotope.

Rhenium-187 is radioactive. A radioactive isotope is one that breaks apart and gives off some form of radiation. The half life of rhenium-187 is about 100,000,000 years. The half life of a radioactive element is the time it takes for half of a sample of the element to break down. For example, of a 100-gram sample of rhenium-187, only half that amount, or 50 grams, would be left after 100,000,000 years. The other 50 grams would have broken down and changed into another isotope.

Extraction

Ores containing rhenium are first roasted, or heated in air, to convert them to rhenium oxide (ReO_3). **Hydrogen** gas is then passed over the rhenium oxide. The hydrogen converts the rhenium oxide to the pure metal:

$$ReO_3 + 3H_2 \rightarrow 3H_2O + Re$$

Rhenium was the last naturally occurring element to be discovered.

Sodium chloride (table salt) crystals.

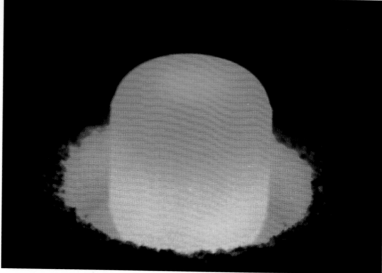

Glowing pellet of plutonium.

Plutonium is used to make nuclear weapons. Here, a computer-enhanced photo shows the
mushroom cloud from the atomic bomb that was dropped over Nagasaki, Japan, on August 9, 1945.

A nine-and-a-half pound button of uranium-235 is held. It will be manufactured into a nuclear weapons component.

Lasers containing thulium are used in satellites that take pictures of the Earth. Here, Intelsat VI floats over the Earth.

A nuclear explosion at sea. The explosion occurs as a result of the uranium-235 isotope undergoing nuclear fission.

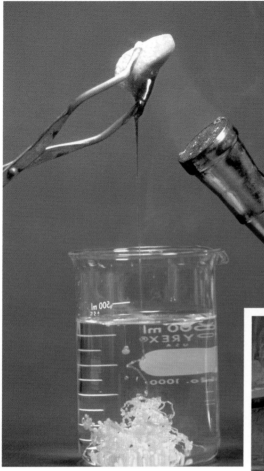

A chemical reaction involving sulfur.

Hot, glowing silver.

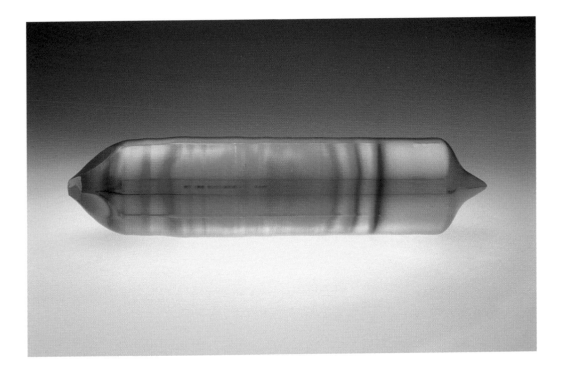

A purified silicon bar.

The Andromeda galaxy. Promethium has been observed in the spectra of some stars in this galaxy.

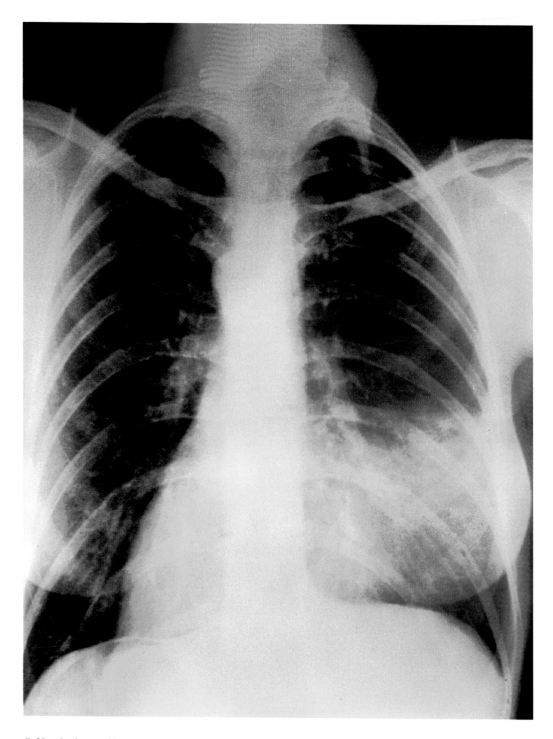

Terbium is often used in X-ray machines. This X-ray shows pneumonia in the lower lobe of the patient's left lung.

Uses

About three-quarters of all rhenium consumed in the United States are used in the manufacture of superalloys. A superalloy is an alloy made of **iron, cobalt,** or **nickel.** It has special properties, such as the ability to withstand high temperatures and attack by oxygen. Superalloys are widely used in making jet engine parts and gas turbine engines.

Alloys containing rhenium also have many other applications. They are used in making devices that control temperatures (like the thermostat in your home), heating elements (like those on an electric stove), vacuum tubes (like those in a television set), electromagnets, electrical contacts, metallic coatings, and thermocouples. A thermocouple is used like a thermometer for measuring very high temperatures.

About a quarter of the rhenium consumed in the United States is used as a catalyst in the petroleum industry. A catalyst is a substance used to speed up or slow down a chemical reaction without undergoing any change itself. Rhenium catalysts are used in the reactions by which natural petroleum is broken down into more useful fragments, such as gasoline, heating oil, and diesel oil.

Compounds

Very few compounds of rhenium have any commercial applications.

Health effects

Complete studies on the health effects of rhenium are not available. For that reason, it should be assumed to be toxic and be handled with caution.

Superalloys are widely used in making jet engine parts and gas turbine engines.

Opposite page:
The heating element found in common household ovens may be made from rhenium alloys.

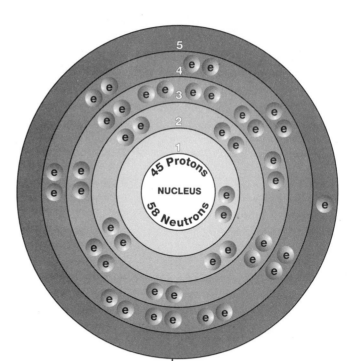

RHODIUM

Overview

Rhodium is considered to be a precious metal. A precious metal is one that is rare and valued. Other precious metals are **gold, silver,** and **platinum.** Rhodium is also classified as a member of the platinum group of metals. The platinum group includes five other metals that often occur together in nature: **ruthenium, palladium, osmium, iridium,** and platinum.

Rhodium falls in the center of the periodic table. The periodic table is a chart that shows how chemical elements are related to one another. Elements in groups 3 through 14 are called the transition elements.

Rhodium was discovered by English chemist and physicist William Hyde Wollaston (1766–1828) in about 1804. He discovered the metal in an ore that apparently came from South America. The rhodium compound he first discovered was a beautiful rose color.

Rhodium is used primarily to make alloys with other metals. These alloys are used for specialized industrial purposes and in jewelry.

SYMBOL
Rh

ATOMIC NUMBER
45

ATOMIC MASS
102.9055

FAMILY
Group 9 (VIIIB)
Transition metal;
platinum group

PRONUNCIATION
RO-dee-um

Alloy a mixture of two or more metals with properties different from those of the individual metals

Catalyst a substance used to speed up or slow down a chemical reaction without undergoing any change itself

Isotopes two or more forms of an element that differ from each other according to their mass number

Precious metal a metal that is rare, desirable, and, therefore, expensive

Radioactive isotope an isotope that breaks apart and gives off some form of radiation

Thermocouple a device for measuring very high temperatures

The first rhodium compound was a beautiful rose color.

Discovery and naming

In the early 1800s, Wollaston was studying an ore of platinum. Although scientists don't know for sure, they believe the platinum ore came from South America. Wollaston analyzed the ore and found that he could produce a beautiful rose-colored compound from it. He showed that the pink compound contained a new element. Wollaston suggested the name rhodium for the new element because of this rose color. The Greek word for rose is *rhodon*.

Physical properties

Rhodium is a silver-white metal. It has a melting point of 1,966°C (3,571°F) and a boiling point of about 4,500°C (8,100°F). Its density is 12.41 grams per cubic centimeter. Two of the metal's special properties are its high electrical and heat conductivity. That means that heat and electricity pass through rhodium very easily.

Chemical properties

Rhodium is a relatively inactive metal. It is not attacked by strong acids. When heated in air, it combines slowly with **oxygen.** It also reacts with **chlorine** or **bromine** when very hot. It does not react with **fluorine,** an element that reacts with nearly every other element.

Occurrence in nature

Rhodium is one of the rarest elements on Earth. Its abundance is estimated to be 0.0001 parts per million. That would place it close to the bottom of the list of elements in terms of abundance. Compounds of rhodium are usually found in combination with platinum and other members of the platinum group. Its most common ores are rhodite, sperrylite, and iridosmine.

Isotopes

Only one naturally occurring isotope of rhodium is known, rhodium-103. Isotopes are two or more forms of an element. Isotopes differ from each other according to their mass number. The number written to the right of the element's name is the mass number. The mass number represents the number of protons plus neutrons in the nucleus of an atom of the element. The number of protons determines the element, but the number of neutrons in the atom of any one element can vary. Each variation is an isotope.

English chemist and physicist William Hyde Wollaston discovered rhodium.

Rhodium also has a number of radioactive isotopes. A radioactive isotope is one that breaks apart and gives off some form of radiation. Radioactive isotopes are produced when very small particles are fired at atoms. These particles stick in the atoms and make them radioactive.

None of the isotopes of rhodium have any commercial or other use.

Extraction

Rhodium is usually obtained as a by-product in the recovery of **platinum** from its ores. Rhodium is separated by a series of chemical and physical reactions from other platinum metals with which it occurs. The mixture of metals is treated with various acids and other chemicals that dissolve some metals, but not others. Rhenium is one of the first metals to be removed from such a mixture.

The cost of pure rhodium was $25 per gram ($600 per troy ounce) in 1997. It cost approximately ten times that in 1991.

Uses

Most of the rhodium metal sold in the United States is used to make alloys. An alloy is made by melting and mixing two or more metals. The mixture has properties different from those of the individual metals. Rhodium is often added to platinum to make an alloy. Rhodium is harder than platinum and has a higher melting point. So the alloy is a better material than pure platinum.

Most rhodium alloys are used for industrial or research purposes, such as laboratory equipment and thermocouples. A thermocouple is a device for measuring very high temperatures. Rhodium alloys are also used to coat mirrors and in searchlights because they reflect light very well.

Compounds

Compounds of rhodium are used as catalysts. A catalyst is a substance used to speed up or slow down a chemical reaction without undergoing any change itself.

Health effects

There are no studies of the health effects from rhodium or its common compounds. Elements without information about toxicity are usually treated as if they are poisonous.

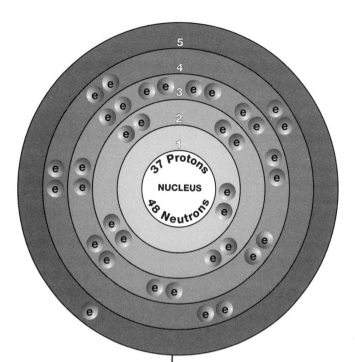

RUBIDIUM

Overview

Rubidium is a soft, silvery metal. It is one of the most active chemical elements. Rubidium is a member of the alkali family. The alkali family consists of elements in Group 1 (IA) of the periodic table. The periodic table is a chart that shows how chemical elements are related to one another. Other Group 1 (IA) elements include **lithium, sodium, potassium, cesium,** and **francium.** Rubidium was discovered in 1861 by German chemists Robert Bunsen (1811–99) and Gustav Kirchhoff (1824–87).

Rubidium is used to make atomic clocks. An atomic clock is a device for keeping very exact time. A radioactive isotope of rubidium is also used to measure the age of very old objects. In general, however, rubidium and its compounds have few practical uses.

Discovery and naming

Rubidium is one of four elements discovered by spectroscopy. Spectroscopy is the process of analyzing the light produced when an element is heated. Every element produces a very specific series of colored lines called a spectrum (plural: *spectra*).

SYMBOL
Rb

ATOMIC NUMBER
37

ATOMIC MASS
85.4678

FAMILY
Group 1 (IA)
Alkali metal

PRONUNCIATION
roo-BID-ee-um

Spectroscopy is a very useful technique for chemists. Sometimes only a very small amount of an element in a sample can be tested. And that amount may be too small to see or weigh easily. But the element can still be detected by heating the sample. The element will give off its characteristic line spectrum. The line spectrum shows that the element is present.

Bunsen and Kirchhoff used a spectroscope to find rubidium in a mineral called lepidolite. The mineral had been discovered in the 1790s by a Jesuit priest, Abbé Nicolaus Poda of Neuhaus, Germany (1723?–98). When Bunsen and Kirchhoff heated a sample of lepidolite, they found two new lines in the spectrum. This is what they reported:

> The magnificent dark red color of these new rays of the new alkali metal led us to give this element the name rubidium and the symbol Rb from *rubidus,* which, with the ancients, served to designate the deepest red.

(*See* sidebar on Bunsen in the **cesium** entry in Volume 1.)

Physical properties
Rubidium is a soft, silvery metal. It has a melting point of 39°C (102°F) and a boiling point of 688°C (1,270°F). Its density is 1.532 grams per cubic centimeter.

Chemical properties
Rubidium is one of the most active elements. It catches fire when exposed to **oxygen** in the air. For that reason, it must be stored completely submerged in kerosene. Rubidium also reacts vigorously with water. It produces **hydrogen** gas which catches fire and burns. Rubidium combines violently with the halogens (**fluorine, chlorine, bromine,** and **iodine).**

Occurrence in nature
Rubidium is a relatively abundant element at about 35 to 75 parts per million. This makes it about as abundant as **nickel, chromium, zinc,** and **copper.**

The most common ores of rubidium are lepidolite, carnallite, and pollucite. Rubidium is also found in seawater and in mineral springs.

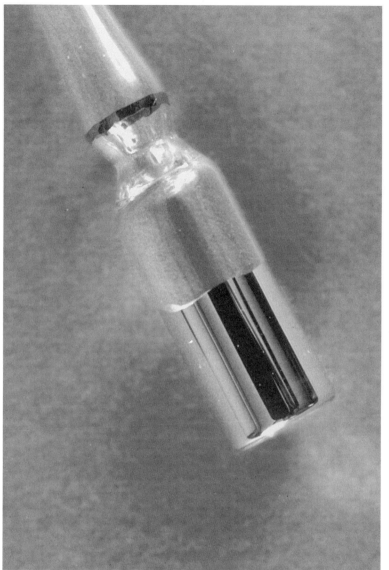

The chemists who discovered rubidium said it had a "magnificent dark red color." The name came from *rubidus,* which was a word once used to refer to "the deepest red."

Sample of rubidium stored in hydrogen.

Isotopes

Two naturally occurring isotopes of rubidium exist: rubidium-85 and rubidium-87. Isotopes are two or more forms of an element. Isotopes differ from each other according to their mass number. The number written to the right of the element's name is the mass number. The mass number represents the number of protons plus neutrons in the nucleus of an atom of the element. The number of protons determines the element, but the number of neutrons in the atom of any one element can vary. Each variation is an isotope.

Rubidium-87 is a radioactive isotope. A radioactive isotope is one that breaks apart and gives off some form of radiation. Some radioactive isotopes occur naturally. Others can be produced artificially by firing very small particles at atoms. These particles stick in the atoms and make them radioactive. About 16 artificial radioactive isotopes of rubidium have also been made.

Rubidium-87 is used to estimate the age of very old rocks. Many kinds of rocks contain two rubidium isotopes, rubidium-85 and rubidium-87. When rubidium-87 breaks down in the rock, it changes into a new isotope, **strontium**-87. Any rock that contains rubidium-87 also contains some strontium-87.

But rocks that contain rubidium usually contain strontium as well. One of the isotopes of strontium found in these rocks is strontium-86. It is not radioactive.

Consider a rock that contains both rubidium and strontium. It will contain two isotopes of strontium. One is naturally-occurring strontium-86, and the other is radioactive strontium-87, which is produced when the rock's rubidium-87 breaks down.

The amount of strontium-87 in the rock depends on how long it has been there. The longer the rock has been in place, the longer rubidium-87 has had to break down and the longer strontium-87 has had to form. But the amount of strontium-86 does not change. It is not produced by rubidium-87.

To determine the age of a rock, then, scientists measure the amount of strontium-87 compared to the amount of strontium-86. The higher the ratio of strontium-87 to strontium-86, the longer the rock has been in existence. This method for measuring the age of rocks has been used to measure the age of the Earth and the age of meteorites.

Extraction

A common method for producing rubidium is to pass an electrical current through molten (melted) rubidium chloride:

$$2RbCl \xrightarrow{\text{electric current}} 2Rb + Cl_2$$

A radioactive isotope of rubidium, rubidium-87 is used to estimate the age of very old rocks.

Uses and compounds

There are relatively few commercial uses for rubidium or its compounds. Rubidium is used to make atomic clocks. But these clocks are used only for very specialized purposes where very precise time-keeping is important. Rubidium is also used to make photocells. A photocell is a device for converting light energy into electrical energy. But other members of the alkali family are still preferred for this application.

One application of photocells is in motion detectors for security alarm systems. A beam of light is emitted by a special device so that it strikes the photocell precisely, producing a tiny electric current. If something or someone "breaks" (interrupts) the beam, the current stops flowing and an alarm sounds.

Health effects

No health effects of rubidium have been recorded. As a safety measure, chemists treat the metal as if it were toxic. They handle it with caution.

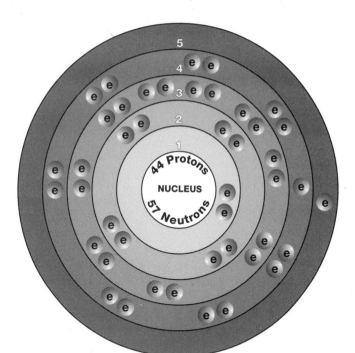

Inside the nucleus diagram:
1
2
3
4
5
44 Protons
NUCLEUS
57 Neutrons

RUTHENIUM

Overview

Ruthenium belongs in the platinum group of metals. The elements in this group are named after the best known member of the group, **platinum.** The group is found in the middle of the periodic table, in Groups 8, 9, and 10, and Rows 5 and 6. The periodic table is a chart that shows how chemical elements are related to one another. The platinum metals tend to be somewhat rare and valuable. They are also called precious metals. The platinum metals also tend to have bright, shiny surfaces and high melting points, boiling points, and densities.

Credit for the discovery of ruthenium is often given to Polish chemist Jedrzej Sniadecki (1768–1838). Sniadecki announced the discovery of the element in 1808. He suggested the name vestium for the element, after the asteroid Vesta. Other chemists were not able to confirm Sniadecki's work, however. As a result, the element was rediscovered twice more in later years.

The primary uses of ruthenium are in alloys and as catalysts for industrial processes.

Discovery and naming

Sniadecki discovered element 44 in 1808 while working with platinum ores from South America. After he published his

SYMBOL
Ru

ATOMIC NUMBER
44

ATOMIC MASS
101.07

FAMILY
Group 8 (VIIIB)
Transition metal;
platinum group

PRONUNCIATION
roo-THEE-nee-um

results, other chemists tried to find the element as well. They were unsuccessful. Sniadecki became discouraged, dropped his claims of discovery, and did no further research on the element.

About twenty years later, the discovery of element 44 was announced again. This time, the discoverer was Russian chemist Gottfried W. Osann. Once more, other chemists could not repeat Osann's results. There was disagreement as to whether the element had been found.

Finally, in 1844, Russian chemist Carl Ernst Claus (also known in Russian as Karl Karlovich Klaus; 1796–1864) gave positive proof of a new element in platinum ores. Many authorities now call Claus the discoverer of the element. Claus suggested calling the element ruthenium, after the ancient name of Russia, *Ruthenia*. Osann had suggested that name as well.

Physical properties

Ruthenium is a hard, silvery-white metal with a shiny surface. Its melting point is about 2,300 to 2,450°C (4,200 to 4,400°F) and its boiling point is about 3,900 to 4,150°C (7,100 to 7,500°F). Its density is 12.41 grams per cubic centimeter.

Chemical properties

Ruthenium metal is relatively unreactive. It does not dissolve in most acids or in aqua regia. Aqua regia is a mixture of hydrochloric and nitric acids. It often reacts with materials that do not react with either acid separately. Ruthenium does not react with **oxygen** at room temperatures either. At higher temperatures, however, it does combine with oxygen.

Occurrence in nature

Ruthenium is one of the rarest elements in the Earth's crust. Its abundance is estimated at about 0.0004 parts per million. This makes it one of six least abundant elements on Earth.

Like other members of the platinum family, ruthenium occurs in platinum ores. It is obtained from those ores and from the mineral osmiridium by purification of the natural material.

Isotopes

Seven isotopes of ruthenium are known. Isotopes are two or more forms of an element. Isotopes differ from each other

WORDS TO KNOW

Alloy a mixture of two or more metals with properties different from those of the individual metals

Aqua regia a mixture of hydrochloric and nitric acids

Catalyst a substance used to speed up or slow down a chemical reaction without undergoing any change itself

Isotopes two or more forms of an element that differ from each other according to their mass number

Periodic table a chart that shows how chemical elements are related to each other

Radioactivity having a tendency to break apart and give off some form of radiation

according to their mass number. The number written to the right of the element's name is the mass number. The mass number represents the number of protons plus neutrons in the nucleus of an atom of the element. The number of protons determines the element, but the number of neutrons in the atom of any one element can vary. Each variation is an isotope.

At least nine radioactive isotopes of ruthenium are also known. A radioactive isotope is one that breaks apart and gives off some form of radiation. Radioactive isotopes are produced when very small particles are fired at atoms. These particles stick in the atoms and make them radioactive.

Ruthenium-106 is used for medical purposes. When ruthenium-106 breaks down, it gives off a form of radiation called beta rays. These beta rays act somewhat like X rays. They attack and kill cancer cells. As an example, ruthenium-106 has been used to treat certain forms of eye cancer.

Extraction

Ruthenium is obtained by separating it from other platinum metals, such as platinum, **palladium,** and **osmium,** with which it occurs. These metals are usually obtained as by-products during the refining of **nickel** metal. They are then separated from each other by a series of chemical reactions.

Uses

One important use of ruthenium is in the manufacture of alloys. An alloy is made by melting and mixing two or more metals. The mixture has properties different from those of the individual metals. Ruthenium adds two properties to an alloy. First, it makes the alloy hard. Second, it makes the alloy resistant to attack by oxygen and other materials.

Ruthenium is most often combined with platinum or **palladium** in alloys. Electrical contacts, devices for measuring very high and very low temperatures, and medical instruments are often made from ruthenium alloys. Some types of jewelry are also made from ruthenium, though it is very expensive.

Ruthenium can also be alloyed with other metals. It is sometimes added to **titanium** to make that metal more resistant to corrosion (rusting). Only 0.1 percent of ruthenium in titanium makes titanium a hundred times more corrosion resistant.

"Ruthenium" comes from the ancient name of Russia, Ruthenia.

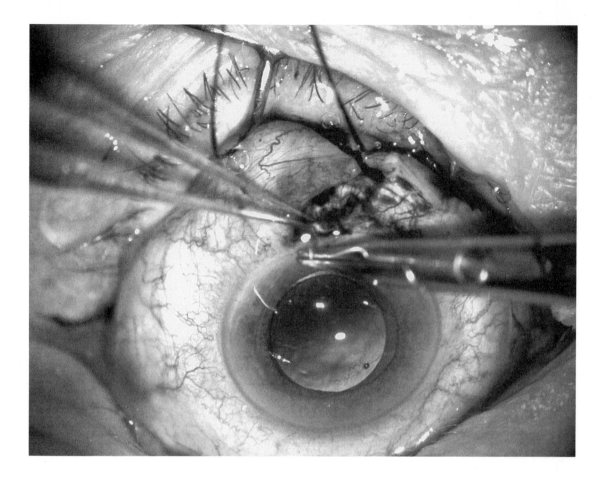

Ruthenium is often combined with platinum or palladium in alloys to make medical instruments. Here, cataract surgery is being performed.

A second important use of ruthenium is in catalysts. A catalyst is a substance used to speed up or slow down a chemical reaction without undergoing any change itself. Ruthenium catalysts may provide a way of changing light energy into electrical energy. The process is similar to photosynthesis, in which green plants change sunlight into stored chemical energy.

Compounds

No compounds of ruthenium have any important commercial application.

Health effects

Ruthenium tetroxide (RuO_4) is very dangerous. Its fumes are irritating to the skin, eyes, and respiratory tract (mouth, throat, and lungs).

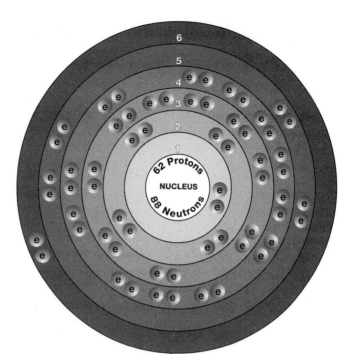

SAMARIUM

Overview

Samarium is one of the rare earth elements found in Row 6 of the periodic table. The periodic table is a chart that shows how chemical elements are related to each other. The rare earth metals are not really very rare in the Earth's surface. The name comes from the fact that these elements are very difficult to separate from each other. For a long time, chemists knew very little about the individual elements. A more correct name for the rare earth elements is the lanthanide series. It is named after the element **lanthanum,** a transition metal also often considered a lanthanide.

Samarium looks and behaves like most other metals, but it has relatively few uses. One of the most important is in the manufacture of very powerful magnets. Compounds of samarium are also used to color glass and in television tubes.

Discovery and naming

The study of chemical elements during the nineteenth century was frustrating. Each time a new element was announced, questions were immediately raised. Was the element really a new element? Or was it a mixture of two or more new elements?

SYMBOL
Sm

ATOMIC NUMBER
62

ATOMIC MASS
150.4

FAMILY
Lanthanide
(rare earth metal)

PRONUNCIATION
suh-MARE-ee-um

The discovery of samarium grew out of this kind of frustration. In 1880, French chemist Paul-Émile Lecoq de Boisbaudran (1838–1912) was studying a substance known as didymium. Earlier chemists believed didymium might be a new element. Boisbaudran said that at least two new elements were present in didymium.

At nearly the same time, French chemist Jean-Charles-Galissard de Marignac (1817–94) was also studying didymium. He was able to separate didymium into two parts, which he called didymium and samarium. He announced that samarium was a new element.

Marignac's research appeared to be satisfactory for nearly twenty years. Then, another French chemist, Eugène-Anatole Demarçay (1852–1904), found that samarium could itself be broken into two parts. He called the new elements samarium and **europium.** Because of this long history, credit for the discovery of samarium is usually given to Boisbaudran, Marignac, Demarçay, or to all three chemists.

The name samarium was taken from a mineral in which it occurs, samarskite. The name of the mineral, in turn, comes from the last name of a Russian mine official, Colonel Samarski.

Physical properties

Samarium is a yellowish metal with a melting point of 1,072°C (1,962°F) and a boiling point of about 1,900°C (3,450°F). Its density is 7.53 grams per cubic centimeter. Samarium is the hardest and most brittle of the rare earth elements.

Chemical properties

Samarium is a fairly reactive metal. It tends to combine with many other substances under relatively mild conditions. For example, it reacts with water to release **hydrogen** gas. It also combines easily with **oxygen** and will ignite (catch fire) at about 150°C (300°F).

Occurrence in nature

As with other rare earth elements, the primary sources of samarium are the mineral monazite and bastnasite. It is also found in samarskite, cerite, orthite, ytterbite, and fluorspar.

WORDS TO KNOW

Isotopes two or more forms of an element that differ from each other according to their mass number

Lanthanides the elements in the periodic table with atomic numbers 58 through 71

Laser a device for making very intense light of one very specific color that is intensified many times over

Radioactive isotope an isotope that breaks apart and gives off some form of radiation

Rare earth elements *see* **Lanthanides**

Toxic poisonous

Sing Praises, *by American sculptor Denise Ward-Brown. Samarium is used to color glass, such as that shown in the book-like part of Ward-Brown's work.*

Samarium is regarded as a relatively abundant lanthanide. It occurs to the extent of about 4.5 to 7 parts per million in the Earth's crust. That makes it about as common as **boron** and two other lanthanides, **thulium** and **gadolinium.**

Isotopes
There are seven naturally occurring isotopes of samarium, samarium-144, samarium-147, samarium-148, samarium-149, samarium-150, samarium-152, and samarium-154. Isotopes are two or more forms of an element. Isotopes differ from each

other according to their mass number. The number written to the right of the element's name is the mass number. The mass number represents the number of protons plus neutrons in the nucleus of an atom of the element. The number of protons determines the element, but the number of neutrons in the atom of any one element can vary. Each variation is an isotope.

Three of samarium's naturally occurring isotopes are radioactive—samarium-147, samarium-148, and samarium-149. A radioactive isotope is one that breaks apart and gives off some form of radiation.

One radioactive isotope of samarium, samarium-153, is used in medicine. Patients with bone cancer often have very severe pain. The isotope samarium-153, can help relieve that pain. It in injected in the form of a drug known as Quadramet. Quadramet was approved by the U.S. Food and Drug Administration (FDA) for this purpose in March 1997.

Extraction

Samarium can be obtained by heating samarium oxide (Sm_2O_3) with **barium** or lanthanum metal:

$$2Ba + Sm_2O_3 \rightarrow Ba_2O_3 + 2Sm$$

Uses

Samarium has some uses similar to those of other rare earth elements. For example, it can be added to glass for color or special optical (light) properties. It is also used to make lasers for special applications. A laser is a device for producing very bright light of a single color. The color produced by the laser depends on the elements it contains.

One of the most important uses of samarium is in the manufacture of very powerful magnets. Samarium is combined with the metal **cobalt** to make samarium-cobalt, or SmCo, magnets. They are among the strongest magnets known. They also have other desirable properties. For example, they can be used at high temperatures and do not react easily with substances around them. SmCo magnets are widely used in motors, such as those used to power specialized kinds of airplanes.

Compounds

The only compound of samarium with any commercial applications is samarium oxide (Sm_2O_3). This compound is used in the manufacture of special kinds of glass, as a catalyst in the man-ufacture of ethanol (ethyl alcohol), and in nuclear power plants as a neutron absorber.

Health effects

The health effects of samarium are not well studied. In such a case, chemists treat the element as toxic and handle it with great caution.

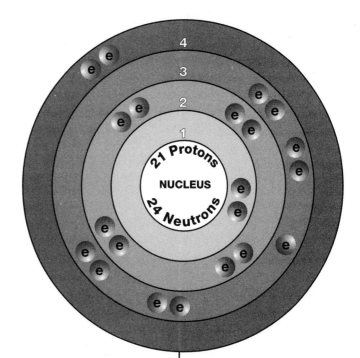

SCANDIUM

Overview

The existence of scandium was predicted nearly ten years before it was actually discovered. The prediction was made by Russian chemist Dmitri Mendeleev (1834–1907). Mendeleev developed the periodic table based on his periodic law. The periodic table is a chart that shows how chemical elements are related to one another. The table originally had a number of empty boxes for elements that had not been discovered. Chemists were able to search for these elements based on the properties of the elements around the empty boxes. Scandium was found in 1879 by Swedish chemist Lars Nilson (1840–99). It is a transition metal, appearing in Group 3 (IIIB).

Scandium is a moderately abundant element. However, it tends to be spread out throughout the earth rather than concentrated in a few places. This makes it difficult to isolate. In fact, scandium is classified as a rare earth element. Rare earth elements are not really "rare." However, they are difficult to extract from the earth. They are also difficult to separate from each other.

Scandium has few commercial uses. It is sometimes combined with other metals to make alloys. An alloy is made by melting

SYMBOL
Sc

ATOMIC NUMBER
21

ATOMIC MASS
44.9559

FAMILY
Group 3 (IIIB)
Transition metal

PRONUNCIATION
SCAN-dee-um

and mixing two or more metals. The mixture has properties different from those of the individual metals. Scandium alloys are being used more in various kinds of sporting equipment and in other applications.

Discovery and naming

In 1869, Mendeleev made one of the great discoveries in the history of chemistry, the periodic law. The periodic law shows how the chemical elements are related to each other. The most common way of representing the periodic law is in a chart called the periodic table.

Mendeleev's original periodic table contained only about 60 elements. That was the total number of elements known in 1869. When he drew his first periodic table, Mendeleev found some empty places. What did those empty places mean?

Mendeleev made a prediction that the empty places in the periodic table stood for elements that had not yet been discovered. He said one could tell what those elements are going to be like by examining their position in the periodic table. For example, element number 21 would be like **boron,** Mendeleev predicted. Boron was the element above number 21 in Mendeleev's chart. He called the missing element (number 21) ekaboron, or "similar to boron."

Chemists were fascinated by Mendeleev's prediction. Could he really tell them how to look for a new element? And could he tell them what that element would be like?

One of the chemists who took up the challenge was Nilson. Nilson analyzed two minerals known as gadolinite and euxenite, in search of the missing element. By 1879, he announced the discovery of "ekaboron." He suggested the name scandium, in honor of Scandinavia, the region in which Nilson' homeland of Sweden is located. (See accompanying sidebar on Nilson.)

Nilson's discovery was very important in chemistry. It showed that Mendeleev's periodic law was correct. The law *did* show how elements are related to each other. It *could* be used to describe elements that had not even been discovered!

The substance discovered by Nilson was not pure scandium metal, but a compound of scandium and **oxygen**—scandium

WORDS TO KNOW

Alloy a mixture of two or more metals with properties different from those of the individual metals

Isotopes two or more forms of an element that differ from each other according to their mass number

Lanthanides the elements in the periodic table with atomic numbers 58 through 71

Periodic table a chart that shows how the chemical elements are related to each other

Radioactive isotope an isotope that breaks apart and gives off some form of radiation

Rare earth elements *See* **Lanthanides**

Lars Nilson | Swedish chemist

Lars Nilson was born in the Swedish town of Östergötland on May 27, 1840. He entered the University of Upsala at the age of 19, intending to study biology, chemistry, and geology. He found university work difficult because he was in very poor health. He often suffered from bleeding in the lungs.

Yet, he persevered and was ready to receive his doctoral degree in 1865. Then he received word that his father was seriously ill. Instead of finishing his university work, he returned home. He took charge of the farm and helped his sick father for many months. At the end of that time, he made a surprising discovery. His illness had disappeared. He was healthy enough to return to Upsala and earn his degree.

In 1879, Nilson made the discovery for which he is most famous. He was studying a mineral known as erbia. The mineral was a complex mixture of many elements. Many chemists throughout Europe were trying to find out exactly what elements were present in erbia.

Nilson found a new element in erbia that no one had yet seen. He was surprised to discover that the element had already been predicted. Russian chemist Dmitri Mendeleev had discovered the periodic law only ten years earlier. Mendeleev had used the periodic law to predict the existence of three elements that had not yet been discovered. One of these elements exactly matched the element found by Nilson. Nilson named the element scandium in honor of his native region, Scandinavia.

oxide (Sc_2O_3). It is quite difficult to produce pure scandium metal from scandium oxide. In fact, it was not until 1937 that the metal was isolated. Then, it was another twenty years before a large sample (weighing one pound) was produced. Today, companies that use scandium often buy the oxide rather than the pure metal. The oxide costs several thousand dollars per kilogram. By comparison, the pure metal costs a few hundred thousand dollars per kilogram.

Physical properties

Scandium metal is a silvery-white solid with a slight pink or yellow tint when exposed to air. It has a melting point of 1,538°C (2,800°F) and a boiling point of about 2,700°C (4,900°F). Its density is 2.99 grams per cubic centimeter.

Chemical properties

Scandium is similar to the rare earth elements chemically. It reacts readily with acids, but does not react easily with oxygen in the air.

Occurrence in nature

The abundance of scandium is thought to be about 5 to 6 parts per million in the Earth's crust. Interestingly, the element seems to be much more abundant in the sun and some stars than it is on Earth.

Scandium is thought to occur in more than 800 different minerals. Its most important ores are the minerals thortveitite and wolframite. It is also found in minerals containing other rare earth elements, such as monazite, bastnasite, and gadolinite.

In the United States, scandium is obtained from the waste products of other mining operations. Some scandium comes from the mining of fluorite at Crystal Mountain, Montana, and some from the mining of **tantalum** in Muskogee, Oklahoma. The actual amount of scandium produced in the United States is not announced. It is regarded as a trade secret in the industry.

Scandium is thought to occur in more than 800 different minerals.

Isotopes

Only one naturally occurring isotope of scandium is known, scandium-45. Isotopes are two or more forms of an element. Isotopes differ from each other according to their mass number. The number written to the right of the element's name is the mass number. The mass number represents the number of protons plus neutrons in the nucleus of an atom of the element. The number of protons determines the element, but the number of neutrons in the atom of any one element can vary. Each variation is an isotope.

About 10 radioactive isotopes of scandium are known also. A radioactive isotope is one that breaks apart and gives off some form of radiation. Radioactive isotopes are produced when very small particles are fired at atoms. These particles stick in the atoms and make them radioactive.

There are no commercial uses for any radioactive isotope of scandium.

Extraction

Pure scandium metal can be made by reacting scandium fluoride (ScF_3) with another active metal, such as **calcium** or **zinc**:

$$3Ca + 2ScF_3 \rightarrow 3CaF_2 + 2Sc$$

Uses

There are relatively few commercial uses for scandium or its compounds. It is sometimes used to make alloys for special purposes. Scandium metal is lighter than most other metals. It is also resistant to corrosion (rusting) and has a high melting point. These properties make scandium alloys especially desirable for use in sporting equipment, such as baseball bats, lacrosse sticks, and bicycle frames. These alloys may also have some applications in the aerospace industry. These applications are not yet well developed, however, because of the high cost of the metal.

Scandium alloys are also used in specialized lamps. The presence of scandium produces light that is very similar to that of natural sunlight.

Compounds

None of the compounds of scandium has any important commercial use.

Scandium alloys are used in bicycle frames.

Scandium

Health effects

As with the rare earth elements, little is known about the health effects of scandium. In such cases, the best policy is to handle the metal very carefully.

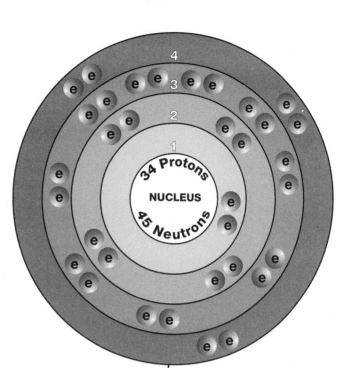

SELENIUM

Overview

Selenium is a member of the chalcogen family. The chalcogens are elements in Group 16 (VIA) of the periodic table. The periodic table is a chart that shows how chemical elements are related to one another. Other chalcogens are **oxygen, sulfur, tellurium,** and **polonium.** The name chalcogen comes from the Greek word *chalkos,* meaning "ore." The first two members of the family, oxygen and sulfur, are found in most ores.

Selenium is a metalloid. A metalloid is an element that has some characteristics of a metal and some of a non-metal.

Selenium and tellurium are often associated with each other. They tend to occur together in the Earth and have somewhat similar properties. They have many uses in common. In recent years, some important new uses have been found for selenium. It is now used in the manufacture of plain paper photocopiers and laser printers, in photovoltaic cells that convert sunlight into electricity, and in X-ray systems for medical applications.

Discovery and naming

Selenium was discovered in 1818 by Swedish chemists Jöns Jakob Berzelius (1779–1848) and J. G. Gahn (1745–1818). The

SYMBOL
Se

ATOMIC NUMBER
34

ATOMIC MASS
78.96

FAMILY
Group 16 (VIA)
Chalcogen

PRONUNCIATION
suh-LEE–nee-um

Allotropes forms of an element with different physical and chemical properties

Alloy a mixture of two or more metals with properties different from those of the individual metals

Amorphous without crystalline shape

Chalcogen elements in Group 16 (VIA) of the periodic table

Isotopes two or more forms of an element that differ from each other according to their mass number

Metalloid an element that acts like a metal sometimes and like a non-metal other times

Periodic table a chart that shows how chemical elements are related to each other

Radioactive isotope an isotope that breaks apart and gives off some form of radiation

Selenium from the Greek word for moon, *selene.*

men were studying the chemicals used in making sulfuric acid at a plant where they had just become part-owners. Among these chemicals they found a material that they thought was the element tellurium. Tellurium had been discovered some 30 years earlier, mixed with some gold deposits in Hungary.

Tellurium is a rare element. Berzelius decided to study the sample more carefully. He took it back to his laboratory in Stockholm. There, he found that he and Gahn had been mistaken. The substance was similar to tellurium, but it also had different properties. They realized they had found a new element. Berzelius suggested naming the element selenium, from the Greek word *selene,* for "moon." The name seemed a good choice because the element tellurium is named after the Latin word *tellus* for "Earth." Just as the Earth and the Moon go together, so do tellurium and selenium.

Physical properties

Selenium exists in a number of allotropic forms. Allotropes are forms of an element with different physical and chemical properties. One allotrope of selenium is an amorphous red powder. Amorphous means "without crystalline shape." A lump of clay is an example of an amorphous material. A second allotrope of selenium has a bluish, metallic appearance. A number of other allotropes have properties somewhere between these two forms.

The amorphous forms of selenium do not have specific melting points. Instead, they gradually become softer as they are heated. They may also change from one color and texture to another.

The crystalline (metallic) form of selenium has a melting point of 217°C (423°F) and a boiling point of 685°C (1,260°F). Its density is 4.5 grams per cubic centimeter.

Some of the most important physical characteristics of selenium are its electrical properties. For example, selenium is a semiconductor. A semiconductor is a substance that conducts an electric current better than non-conductors, but not as well as conductors. Semiconductors have many very important applications today in the electronics industry. Selenium is often used in the manufacture of transistors for computers, cellular phones, and hand-held electronic games.

Selenium pellets.

Selenium is also a photoconductor, a material that changes light energy into electrical energy. Furthermore, it becomes better at making this conversion as the light intensity or brightness increases.

Chemical properties

Selenium is a fairly reactive element. It combines easily with **hydrogen, fluorine, chlorine,** and **bromine.** It reacts with nitric and sulfuric acids. It also combines with a number of metals to form compounds called selenides. An example is **magnesium** selenide (MgSe). One of its interesting reactions is with oxygen. It burns in oxygen with a bright blue flame to form selenium dioxide (SeO_2). Selenium dioxide has a characteristic odor of rotten horseradish.

Occurrence in nature

Selenium is a very rare element. Scientists estimate its abundance at about 0.05 to 0.09 parts per million. It ranks among the

Selenium and tellurium are often associated with each other. They tend to occur together in the Earth and have somewhat similar properties.

25 least common elements in the Earth's crust. It is widely distributed throughout the crust. There is no ore from which it can be mined with profit. Instead, it is obtained as a by-product of mining other metals. It is now produced primarily from **copper, iron,** and **lead** ores. The major producers of selenium in the world are Japan, Canada, Belgium, the United States, and Germany.

Isotopes

There are six naturally occurring isotopes of selenium, selenium-74, selenium-76, selenium-77, selenium-78, selenium-80, and selenium-82. Isotopes are two or more forms of an element. Isotopes differ from each other according to their mass number. The number written to the right of the element's name is the mass number. The mass number represents the number of protons plus neutrons in the nucleus of an atom of the element. The number of protons determines the element, but the number of neutrons in the atom of any one element can vary. Each variation is an isotope.

About a dozen radioactive isotopes of selenium are known also. A radioactive isotope is one that breaks apart and gives off some form of radiation. Radioactive isotopes are produced when very small particles are fired at atoms. These particles stick in the atoms and make them radioactive.

Only one radioactive isotope of selenium is used commercially, selenium-75. This isotope is used to study the function of two organs in the body, the pancreas and the parathyroid gland. (The pancreas helps with digestion and the parathyroid gland releases hormones.) The radioactive selenium is injected into the blood stream. It then goes primarily to one or both of these two organs. The isotope gives off radiation when it reaches these organs. A technician can tell whether the organs are functioning properly by the amount and location of radiation given off.

Extraction

Selenium is obtained as a by-product from other industrial processes. For example, when copper is refined, small amounts of selenium are produced as by-products. This selenium can be removed from the copper-refining process and purified. Selenium is also obtained as a secondary product during the manufacture of sulfuric acid.

The miracle of copying

We sometimes forget what an amazing step forward the invention of the copy machine was. Hundreds of years ago, making a copy of a document was a long, difficult process. Some people spent their whole lives making copies of important documents. Each copy was written out by hand. The process was not only dull and monotonous, but it also resulted in many errors.

Even thirty years ago, copying was slow and difficult. For example, carbon paper allowed a person to make one or more copies while writing or typing. But every error had to be corrected on every copy. The copies were often messy and difficult to read.

Mimeograph machines made it possible to reproduce dozens of copies in a few minutes, but required handwritten or typed originals. The final product was printed in purple.

Then came the photocopy machine. Copies could be made by simply placing the original on a glass cover and pushing a button. What goes on inside a copy machine to make this happen?

An essential part of a photocopier is a drum-shaped unit or a wide moving belt. Fine selenium powder is spread on the surface of the drum or the belt. An electric charge is then applied to the selenium.

Another part of the photocopy machine consists of a set of mirrors. When the machine's "Copy" button is pushed, a bright light shines on the page being copied. The light reflects off the white parts of the page. But it is not reflected off the dark parts, such as text or images. The light reflects off the mirrors to the drum or belt.

Selenium is important because when light strikes the charged selenium, the charge disappears. The sections on the drum or belt struck by light have no charge. The sections *not* struck by light continue to have a charge.

Next, a toner is spread out over the surface of the drum or belt. A toner is usually finely-divided carbon. It sticks to the areas that still carry an electric charge. But it does not stick to the selection without a charge.

Finally, a piece of paper is pressed against the drum or belt. The toner sticks to the paper. A blast of heat causes the carbon to melt and stick tightly to the paper. A copy of the original document is produced by the machine.

Uses

The two most important uses of selenium are in glass-making and in electronics. Each accounts for about 30 to 35 percent of all the selenium produced each year. The addition of selenium to glass can have one of two opposite effects. First, it will cancel out the green color that iron compounds usually add to glass. If a colorless glass is desired, a little selenium is added to neutralize the effects of iron. Second, selenium will add its own color—a beautiful ruby red—if that is wanted in a glass product.

Selenium plays a critical role in the photocopying process.

Selenium is also added to glass used in architecture. The selenium reduces the amount of sunlight that gets through the glass.

A growing use of selenium is in electronic products. One of the most important uses is in plain-paper photocopiers and laser printers. The element is also used to make photovoltaic ("solar") cells. When light strikes selenium, it is changed into electricity. A solar cell is a device for capturing the energy of sunlight on tiny pieces of selenium. The sunlight is then changed into electrical energy.

Currently, that process is not very efficient. Too much sunlight is lost without being converted into electricity. More efficient solar cells will be able to make use of all the free sunlight that strikes the planet every day.

About a third of all selenium produced is used as pigments (coloring agents) for paints, plastics, ceramics, and glazes.

Depending on the form of selenium used, the color ranges from deep red to light orange.

Selenium is also used to make alloys. An alloy is made by melting and mixing two or more metals. The mixture has properties different from those of the individual metals. The addition of selenium to a metal makes it more machinable. Machinability means working with a metal: bending, cutting, shaping, turning, and finishing the metal, for example.

About 5 percent of all selenium produced is used in agriculture. It is added to soil or animal feed to provide the low levels of selenium needed by plants and animals.

Compounds
Very few compounds of selenium have any important practical applications. One exception is selenium sulfide (SeS_2). This compound is used to treat seborrhea, or "oily skin." It is sometimes added to shampoos for people with especially oily hair. Another compound, selenium diethyldithiocarbonate ($Se[SC(S)N(C_2H_5)_2]_4$), is used as a vulcanizing ("toughening") agent for rubber products.

Health effects
Selenium has some rather interesting nutritional roles. It is essential in very small amounts for the health of both plants and animals. Animals that do not have enough selenium in their diets may develop weak muscles. But large doses of selenium are dangerous. In some parts of California, for example, selenium has been dissolved out of the soil by irrigation systems. Lakes accumulate unusually high levels of selenium and birds and fish in the area develop health problems.

A serious selenium problem occurred at the Kesterson Reservoir in Northern California. In the late 1970s, scientists found that birds nesting in the reservoir were developing genetic deformities. They traced the problem to high levels of selenium in the water. A large artificial lake was built and the birds were moved to the artificial lake. They were no longer allowed to nest in the dangerous waters of the reservoir.

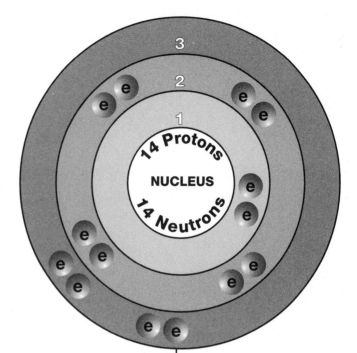

SILICON

Overview

Silicon is a member of Group 14 (IVA) in the periodic table. The periodic table is a chart that shows how chemical elements are related to one another. Silicon is also part of the the carbon family. Other carbon family elements include **carbon, germanium, tin,** and **lead.** Silicon is a metalloid, one of only a very few elements that have characteristics of both metals and non-metals.

Silicon is the second most abundant element in the Earth's crust, exceeded only by **oxygen.** Many rocks and minerals contain silicon. Examples include sand, quartz, clays, flint, amethyst, opal, mica, feldspar, garnet, tourmaline, asbestos, talc, zircon, emerald, and aquamarine. Silicon never occurs as a free element. It is always combined with one or more other elements as a compound.

By the early 1800s, silicon was recognized as an element. But chemists had serious problems preparing pure silicon because it bonds (attaches) tightly to oxygen. It took chemists many years to find out how to separate silicon from oxygen. That task was finally accomplished in 1823 by Swedish chemist Jöns Jakob Berzelius (1779–1848).

SYMBOL
Si

ATOMIC NUMBER
14

ATOMIC MASS
28.0855

FAMILY
Group 14 (IVA)
Carbon

PRONUNCIATION
SIL-i-con

Silicon's most important application is in electronic equipment. Silicon is one of the best materials from which to make transistors and computer chips. The total weight of silicon used for this purpose is relatively small. Much larger amounts are used, for example, to make metal alloys. An alloy is made by melting and mixing two or more metals. The mixture has properties different from those of the individual metals.

Discovery and naming

In one sense, humans have always used silicon. Nearly every naturally occurring rock or mineral contains some silicon. So when ancient peoples built clay huts or sandstone temples, they were using compounds of silicon.

But no one thought about silicon as an element until the nineteenth century. Then, a number of chemists tried to separate silicon from the other elements with which it is combined in the earth. English scientist Sir Humphry Davy (1778–1829) developed a technique for separating elements that tightly bond to each other. He melted these compounds and passed an electric current through them. The technique was successful for producing free or elemental **sodium, potassium, calcium,** and a number of other elements for the first time. But he failed with silicon. (See sidebar on Davy in the calcium entry in Volume 1.)

Berzelius also tried to isolate silicon using a method similar to that of Davy's. He mixed molten (melted) potassium metal with a compound known as potassium silicon fluoride (K_2SiF_6). The result was a new element—silicon.

$$4K + K_2SiF_6 \longrightarrow \text{heat and electricity} \rightarrow 6KF + Si$$

Scottish chemist Thomas Thomson (1773–1852) suggested the name silicon, based on the Latin word for "flint," *silex* (or *silicis*). He added the ending *-on* because the new element was so much like bor*on* and carb*on*. Thus, the new element's name was accepted as silicon.

Some interesting studies were done on silicon over the next few years. German chemist Friedrich Wöhler (1800–82) produced a series of compounds known as silanes. These compounds contain silicon, **hydrogen,** and, sometimes, other ele-

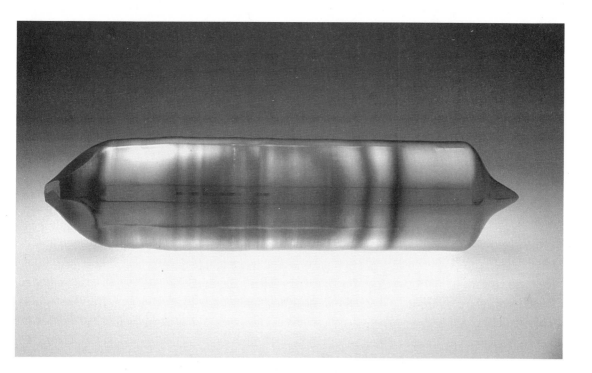

ments. The simplest silane is silicon tetrahydride (SiH_4). This compound is also called silane.

Purified silicon bar.

A group of compounds known as the siloxanes were produced at about the same time. The siloxanes are made up of silicon, oxygen, and an organic group. Organic compounds contain carbon.

Silanes and siloxanes were not discovered in the search for the answer to any practical question. Chemists were just curious about the kinds of compounds they could make with silicon. But many years later, chemists made some interesting discoveries. Both groups of compounds do have some very important practical uses. For example, the compounds known as silicones are a form of the siloxanes.

Physical properties

Silicon is a metalloid, an element with properties of both metals and non-metals. Silicon exists in two allotropic forms. Allotropes are forms of an element with different physical and chemical properties. One allotrope is in the form of shiny, grayish-black, needle-like crystals, or flat plates. The second allotrope has no crystal structure and usually occurs as a brown powder.

The melting point of silicon is 1,410°C (2,570°F) and the boiling point is 2,355°C (4,270°F). Its density is 2.33 grams per cubic centimeter. Silicon has a hardness of about 7 on the Mohs scale. The Mohs scale is a way of expressing the hardness of a material. It runs from 0 (for talc) to 10 (for diamond).

Silicon is a semiconductor. A semiconductor is a substance that conducts an electric current better than a non-conductor—like glass or rubber—but not as well as a conductor—like **copper** or **aluminum.** Semiconductors have important applications in the electronics industry.

Chemical properties

Silicon is a relatively inactive element at room temperature. It does not combine with oxygen or most other elements. Water, steam, and most acids have very little affect on the element. At higher temperatures, however, silicon becomes much more reactive. In the molten (melted) state, for example, it combines with oxygen, **nitrogen, sulfur, phosphorus,** and other elements. It also forms a number of alloys very easily in the molten state.

Occurrence in nature

Silicon is the second must abundant element in the Earth's crust. Its abundance is estimated to be about 27.6 percent of the crust. It ranks second only to oxygen. Some authorities believe that more than 97 percent of the crust is made of rocks that contain compounds of silicon and oxygen.

Silicon has been detected in the Sun and stars. It also occurs in certain types of meteorites known as aerolites or "stony meteorites." Meteorites are rock-like chunks that fall to the Earth's surface from outside the Earth's atmosphere.

Silicon never occurs as a free element in nature. It always occurs as a compound with oxygen, **magnesium,** calcium, phosphorus, or other elements. The most common minerals are those that contain silicon dioxide in one form or another. These are known as silicates.

Isotopes

There are three naturally occurring isotopes of silicon: silicon-28, silicon-29, and silicon-30. Isotopes are two or more forms of an element. Isotopes differ from each other according to their mass number. The number written to the right of the ele-

Silicon has been detected in the Sun and stars. It also occurs in certain types of meteorites.

ment's name is the mass number. The mass number represents the number of protons plus neutrons in the nucleus of an atom of the element. The number of protons determines the element, but the number of neutrons in the atom of any one element can vary. Each variation is an isotope.

Five radioactive isotopes of silicon are known also. A radioactive isotope is one that breaks apart and gives off some form of radiation. Radioactive isotopes are produced when very small particles are fired at atoms. These particles stick in the atoms and make them radioactive.

None of the radioactive isotopes of silicon has any commercial use.

Extraction
Silicon is prepared by heating silicon dioxide with carbon. Carbon replaces the silicon in the compound. The silicon formed is 96 to 98 percent pure.

$$SiO_2 + C \xrightarrow{\text{heat}} CO_2 + Si$$

Many applications of silicon require a very pure product. Methods have been developed to produce silicon that is at least 99.97 percent pure silicon. This form of silicon is called hyperpure silicon.

Uses
Perhaps the best known use of silicon is in electronic devices. Hyperpure silicon is used in transistors and other components of electronic devices. It is also used to make photovoltaic (solar) cells, rectifiers, and parts for computer circuits. A photovoltaic cell is a device that converts sunlight into electrical energy. A rectifier is an electrical device for changing one kind of electric current (alternating current, or AC) into another kind of electric current (direct current, or DC).

The largest single use of silicon, however, is in making alloys. The most important silicon alloys are those made with **iron** and steel, aluminum, and copper. When silicon is produced, in fact, scrap iron and metal is sometimes added to the furnace. As soon as the silicon is produced, it reacts with iron and steel to form ferrosilicon. Ferrosilicon is an alloy of iron or steel and silicon. It is used for two major purposes. First, it can be

Almost without exception, all glass contains silicon dioxide.

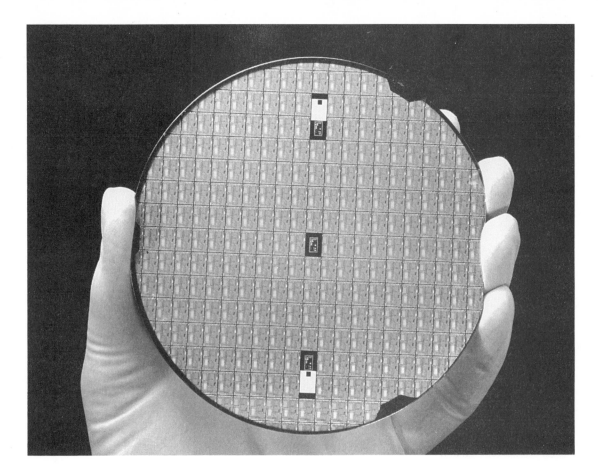

A circular wafer of silicon carrying many individual integrated circuits.

added to steel to improve the strength and toughness of the steel. Second, it can be added during the steel-making process to remove impurities from the steel that is being made.

The aluminum industry uses large amounts of silicon in alloys. These alloys are used to make molds and in the process of welding. Welding is a process by which two metals are joined to each other. Alloys of silicon, aluminum, and magnesium are very resistant to corrosion (rusting). They are often used in the construction of large buildings, bridges, and transportation vehicles such as ships and trains.

Compounds

A number of silicon compounds have important uses. Silicon dioxide (sand) is used in the manufacture of glass, ceramics, abrasives, as a food additive, in water filtration systems, as an insulating material, in cosmetics and pharmaceuticals (drugs), and in the manufacture of paper, rubber, and insecticides. Each

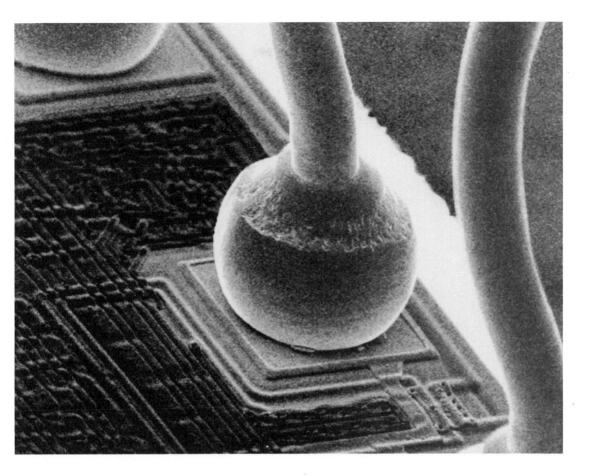

A wire bonded to a silicon chip that houses an integrated circuit. The photo is magnified 280 times.

of these applications could be the subject of a very long dis-cussion in and of itself. For example, humans have made glass for thousands of years. Today, dozens of different kinds of glass are produced, each with special properties and uses. But almost without exception, they all contain silicon dioxide.

Another important compound is silicon carbide (SiC). Silicon carbide is also known as carborundum. It is one of the hardest substances known, with a hardness of about 9.5 on the Mohs scale. Carborundum is widely used as an abrasive, a powdery material used to grind or polish other materials. Carborundum also has refractory properties. A refractory material can with-stand very high temperatures by reflecting heat. Refractory materials are used to line the inside of ovens used to maintain very high temperatures.

Another important silicon group is the silicones. The silicones have an amazing range of uses. These include toys (Silly Putty and Superballs), lubricants, weatherproofing materials, adhe-

sives (glues), foaming agents, brake fluids, cosmetics, polishing agents, electrical insulation, materials to reduce vibration, shields for sensitive equipment, surgical implants, and parts for automobile engines.

Health effects

Information on the health effects of silicon is limited. Some studies show that silicon may be needed in very small amounts by plants and some animals. One study showed, for example, that chickens that did not receive silicon in their diet developed minor health problems. Overall, silicon probably has no positive or negative effects on human health.

However, a serious health problem called silicosis is associated with silicon dioxide (SiO_2). Silicon dioxide occurs in many forms in the earth. Ordinary sand is nearly pure silicon dioxide.

In some industries, sand is ground up into a very fine powder that gets into the air. As workers inhale the dust, it travels through their mouths, down their throats, and into their lungs. Silicon dioxide powder can block the tiny air passages in the lungs through which oxygen and carbon dioxide pass. When this happens, silicosis results.

Silicosis is similar to pneumonia. The person finds it difficult to breathe. The longer one is exposed to silicon dioxide dust, the worst the problem gets. In the worst cases, silicosis results in death because of the inability to breathe properly.

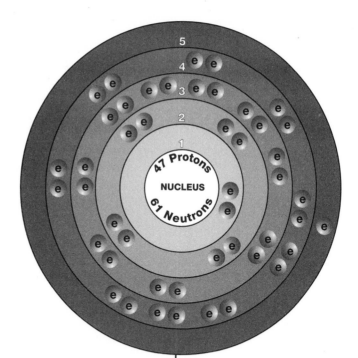

SILVER

Overview

Chemists classify silver as a transition metal. The transition metals are elements between Groups 2 and 13 in the periodic table. The periodic table is a chart that shows how chemical elements are related to one another. More than 40 elements, all metals, fall within the transition metal range.

Silver is also classified as a precious metal. Precious metals are not very abundant in the Earth's crust. They are attractive and not very chemically active. These properties make the metal desirable in jewelry, coins, and art. About a half dozen metals near silver in the periodic table are also precious metals. These include **gold, platinum, palladium, rhodium,** and **iridium.**

Silver has been used by humans for thousands of years. It often occurs as a free element in nature. It can also be extracted from its ores fairly easy. These properties made it easy for early humans to learn about silver.

Today, the most important use of silver is in photography. Three silver compounds used in photography are silver chloride (AgCl), silver bromide (AgBr), and silver iodide (AgI). Silver is also used to make electrical equipment, mirrors, medical and

SYMBOL
Ag

ATOMIC NUMBER
47

ATOMIC MASS
107.868

FAMILY
Group 11 (IB)
Transition metal

PRONUNCIATION
SIL-ver

WORDS TO KNOW

Alloy a mixture of two or more metals with properties different from those of the individual metals

Amalgam an alloy of mercury and at least one other metal

Ductile capable of being drawn into thin wires

Isotopes two or more forms of an element that differ from each other according to their mass number

Electrolysis a process by which a compound is broken down by passing an electric current through it

Malleable capable of being hammered into thin sheets

Periodic table a chart that shows how chemical elements are related to each other

Radioactive isotope an isotope that breaks apart and gives off some form of radiation

Silver plating a process by which a very thin layer of silver metal is laid down on top of another metal

Toxic poisonous

Transition metal an element in Groups 3 through 12 of the periodic table

dental equipment, and jewelry. It is often used to make alloys with gold for some of these applications. An alloy is made by melting and mixing two or more metals. The mixture has properties different from those of the individual metals.

Discovery and naming

Silver was probably first discovered after gold and **copper.** Gold and copper often occur as free elements in nature. They have very distinctive colors, which made it easy for early humans to find these metals.

Silver also occurs as a free metal, but much less often than gold or copper. At some point, humans learned to extract silver from its ores. But that discovery must have occurred very early on in human history. Archaeologists (scientists who study ancient civilizations) have found silver objects dating to about 3400 B.C. in Egypt. Drawings on some of the oldest pyramids show men working with metal, probably extracting silver from its ores.

Other early cultures also used silver. Written records from India describe the metal as far back as about 900 B.C. Silver was in common use in the Americas when Europeans first arrived.

The Bible contains many references to silver. The metal was used as a way of paying for objects. It also decorated temples, palaces, and other important buildings. The Bible also contains sections that describe the manufacture of silver.

The word silver goes back to at least the 12th century, A.D. It seems to have come from an old English word used to describe the metal, *seolfor.* The symbol for silver (Ag), however, comes from its Latin name, *argentum.* The name may have originated from the Greek word *argos,* meaning "shiny" or "white."

Physical properties

Silver is a soft, white metal with a shiny surface. It is the most ductile and most malleable metal. Ductile means capable of being drawn into thin wires. Malleable means capable of being hammered into thin sheets. Silver has two other unique properties. It conducts heat and electricity better than any other element. It also reflects light very well.

Drawings on some of
the oldest pyramids
show men working with
metal, probably
extracting silver from
its ores.

Hot, glowing silver.

Silver's melting point is 961.5°C (1,762°F) and its boiling
point is about 2,000 to 2,200°C (3,600 to 4,000°F). Its density is 10.49 grams per cubic centimeter.

Chemical properties

Silver is a very inactive metal. It does not react with **oxygen** in
the air under normal circumstances. It does react slowly with
sulfur compounds in the air, however. The product of this
reaction is silver sulfide (Ag_2S), a black compound. The tarnish
that develops over time on silverware and other silver-plated
objects is silver sulfide.

Silver does not react readily with water, acids, or many other compounds. It does not burn except as silver powder.

Occurrence in nature

Silver is a fairly rare element in the Earth's crust. Its abundance is estimated to be about 0.1 parts per million. It is also found in seawater. Its abundance there is thought to be about 0.01 parts per million.

Silver usually occurs in association with other metal ores, especially those of **lead.** The most common silver ores are argentite (Ag_2S); cerargyrite, or "horn silver" ($AgCl$); proustite ($3Ag_2S \bullet As_2S_3$); and pyrargyrite ($Ag_2S \bullet Sb_2S_3$).

The largest producers of silver in the world are Mexico, Peru, the United States, Canada, Poland, Chile, and Australia. In the United States, silver is produced at about 76 mines in 16 states. The largest state producers are Nevada, Idaho, and Arizona. These three states account for about two-thirds of all the silver mined in the United States.

Isotopes

Two naturally occurring isotopes of silver exist: silver-107 and silver-109. Isotopes are two or more forms of an element. Isotopes differ from each other according to their mass number. The number written to the right of the element's name is the mass number. The mass number represents the number of protons plus neutrons in the nucleus of an atom of the element. The number of protons determines the element, but the number of neutrons in the atom of any one element can vary. Each variation is an isotope.

About 16 radioactive isotopes of silver are known also. A radioactive isotope is one that breaks apart and gives off some form of radiation. Radioactive isotopes are produced when very small particles are fired at atoms. These particles stick in the atoms and make them radioactive.

None of the radioactive isotopes of silver has any commercial use.

Extraction

Ores rich in silver disappeared long ago due to mining. Today, silver usually comes from ores that contain very small amounts of the metal. These amounts can range from about a few thou-

The tarnish that develops over time on silverware and other silver-plated objects is silver sulfide.

A small percent of silver produced in the United States is used for coins. The old "Peace" silver dollar, shown here, was minted from 1921 to 1935.

sandths of an ounce per ton of ore to 100 ounces per ton. The metal is most commonly produced as a by-product of mining for other metals. After the primary metal has been removed, the waste often contains small amounts of silver. These wastes are treated with chemicals that react with the silver. The silver can then be extracted by electrolysis. Electrolysis is a process by which a compound is broken down by passing an electric current through it.

Uses and compounds

About 10 percent of silver produced in the United States is used in coins, jewelry, and artwork. One way silver is used is in alloys with gold. Gold is highly desired for coins and jewelry. But it is much too soft to use in its pure form. Adding silver to gold, however, makes an alloy that is much stronger and longer lasting. Most "gold" objects today are actually alloys, often alloys of silver and gold.

Other objects use much more of the silver metal, however. About half of the silver produced in the United States goes

Silver's important role in film

Taking a photograph depends on a simple chemical idea: Light can cause electrons to move around. Here is what that means:

Silver metal will combine with chlorine, bromine, or iodine to form compounds. As an example:

$$2Ag + Cl_2 \rightarrow 2AgCl$$

In this reaction, each silver atom loses one electron to a chlorine atom. The silver atom becomes "one electron short" of what it usually has. The one-electron-short silver atom is called a silver ion.

Photographic film is coated with a thin layer of silver chloride, silver bromide, or silver iodide. That means the film is covered with many silver ions. Silver ions are colorless, so photographic film has no color to it.

What happens when photographic film is exposed to light? Light gives energy to electrons in the photographic film. Some of these electrons find their way back to silver ions, transforming them back to atoms:

$$silver\ ion + electron \rightarrow silver\ atom$$

But silver atoms are not colorless. They are black. So, a photographic film exposed to light turns black at every point where light strikes a silver ion.

In taking a picture, of course, not all of the film gets equal amounts of light. A picture of a person, for example, will have areas that get much more light than others. So some places on the film become very dark, and other places become less dark.

Additional steps are necessary to "develop" photographic film or to produce a picture from it. But the first step in taking a photograph is changing silver ions back to silver atoms with light.

into photographic film. Pure silver is first converted to a compound: silver chloride, silver bromide, or silver iodide. The compound is then used to make photographic film (see accompanying sidebar).

The second most important use of silver is in electrical and electronic equipment. About 20 percent of all silver produced is used for this purpose. Silver is actually the most desirable of all metals for electrical equipment. Electricity flows through silver more easily than it does through any other metal. In most cases, however, metals such as copper or **aluminum** are used because they are less expensive.

But sometimes, an electrical device is so important that cost is not a consideration. For example, electrical devices on spacecraft, satellites, and aircraft must work reliably and efficiently.

The cost of using silver is not as important as it would be in a home appliance. Thus, silver is used for electrical wiring and connections in these devices.

In some cases, silver plating solves a practical problem where the more expensive silver would work best. Silver plating is the process by which a very thin layer of silver metal is laid down on top of another metal. Silver is so malleable that it can be hammered into sheets thinner than a sheet of paper. Silver this thin can be applied to another metal. Then the other metal takes on some of the properties of the silver coating. For example, it may work very well as a reflector because silver is such a good reflector. It does not matter if the second metal is a good reflector or not. The silver coating serves as the reflecting surface in the combination.

About a fifth of all silver produced is used in a variety of other products. For example, it is often used in dental amalgams. An amalgam is an alloy in which mercury is one of the metals used. Silver amalgams work well for filling decayed teeth. They are non-toxic and do not break down or react with other materials very readily. Silver is also used in specialized batteries, including silver-**zinc** and silver-**cadmium** batteries.

Health effects
Silver is a mildly toxic element. When the metal or its compounds get on the skin, they can cause a bluish appearance known as argyria or argyrosis. Breathing in silver dust can have serious long-term health effects also. The highest recommended exposure for silver dust is 0.1 milligrams per cubic meter of air.

Electricity flows through silver more easily than it does through any other metal.

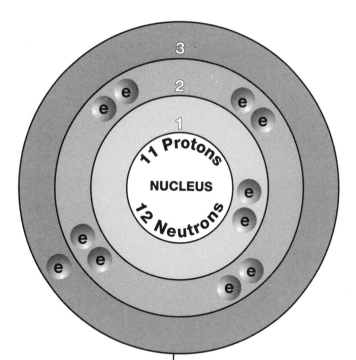

SODIUM

Overview

Most people have never seen sodium metal. But it is almost impossible not to see many compounds of sodium every day. Ordinary table salt, baking soda, baking powder, household lye (such as Drano), soaps and detergents, aspirin and other drugs, and countless other consumer products are sodium products.

Sodium is a member of the alkali metals family. The alkali family consists of elements in Group 1 (IA) of the periodic table. The periodic table is a chart that shows how chemical elements are related to one another. Other Group 1 (IA) elements are **lithium, potassium, rubidium, cesium,** and **francium.** The members of the alkali metals family are among the most active elements.

Compounds of sodium have been known, of course, throughout human history. But sodium metal was not prepared until 1807. The reason is that sodium attaches itself very strongly to other elements. Its compounds are very difficult to break apart. It was not until 1807 that English chemist Sir Humphry Davy (1778–1829) found a way to extract sodium from its compounds. (See sidebar on Davy in the **calcium** entry in Volume 1.)

SYMBOL
Na

ATOMIC NUMBER
11

ATOMIC MASS
22.98977

FAMILY
Group 1 (IA)
Alkali metal

PRONUNCIATION
SO-dee-um

WORDS TO KNOW

Amalgam an alloy of mercury and at least one other metal

Catalyst a substance used to speed up or slow down a chemical reaction without undergoing any change itself

Ductile capable of being drawn into thin wires

Heat exchange medium a material that picks up heat in one place and carries it to another place

Isotopes two or more forms of an element that differ from each other according to their mass number

Malleable capable of being hammered into thin sheets

Periodic table a chart that shows how chemical elements are related to each other

Radioactive isotope an isotope that breaks apart and gives off some form of radiation

Tracer a radioactive isotope whose presence in a system can easily be detected

Sodium metal itself has relatively few uses. It reacts with other substances easily, sometimes explosively. However, many sodium compounds have many uses in industry, medicine, and everyday life.

Discovery and naming

Sodium carbonate, or soda (Na_2CO_3), was probably the sodium compound best known to ancient peoples. It is the most common ore of sodium found in nature.

This explains why glass was one of the first chemical products made by humans. Glass is made by heating sodium carbonate and **calcium** oxide (lime) together. When the mixture cools, it forms the hard, clear, transparent material called glass. Glass was being manufactured on a large scale in Egypt as early as 1370 B.C.

The Egyptians called soda *natron*. Much later, the Romans used a similar name for the compound, *natrium*. These names explain the chemical symbol used for sodium, Na.

The name sodium probably originated from an Arabic word *suda*, meaning "headache." Soda was sometimes used as a cure for headaches among early peoples. The word *suda* also carried over into Latin to become *sodanum*, which also means "headache remedy."

In the early 1800s, Davy found a way to extract a number of active elements from their compounds. Sodium was one of these elements. Davy's method involved melting a compound of the active element, then passing an electric current through the molten (melted) compound. Davy used sodium hydroxide (NaOH) to make sodium.

Physical properties

Sodium is a silvery-white metal with a waxy appearance. It is soft enough to be cut with a knife. The surface is bright and shiny when first cut, but quickly becomes dull as sodium reacts with **oxygen** in the air. A thin film of sodium oxide (Na_2O) forms that hides the metal itself.

Sodium's melting point is 97.82°C (208.1°F) and its boiling point is 881.4°C (1,618°F). Its density is slightly less than that of water, 0.968 grams per cubic centimeter. Sodium is a good conductor of electricity.

Sodium and water aren't friends

Oil and vinegar don't mix. But sodium and water *really* don't mix! Sodium reacts violently with water. The effect is fascinating.

When sodium metal is first placed into water, it floats. But it immediately begins to react with water, releasing hydrogen gas:

$$2Na + 2H_2O \rightarrow 2NaOH + H_2$$

A great deal of energy is released in this reaction. It is enough to set fire to the hydrogen gas. The sodium metal reacts with water. So much heat is released that the sodium melts. It turns into a tiny ball of liquid sodium. At the same time, the sodium releases hydrogen from water. The hydrogen gas catches fire and causes the ball of sodium to go sizzling across the surface of the water.

Chemical properties

Sodium is a very active element. It combines with oxygen at room temperature. When heated, it combines very rapidly, burning with a brilliant golden-yellow flame.

Sodium also reacts violently with water. (See accompanying sidebar.) It is so active that it is normally stored under a liquid with which it does not react. Kerosene or naphtha are liquids commonly used for this purpose.

Sodium also reacts with most other elements and with many compounds. It reacts with acids to produce hydrogen gas. It also dissolves in **mercury** to form a sodium amalgam. An amalgam is an alloy of mercury and at least one other metal.

Sodium reacts violently with water.

Occurrence in nature

Sodium never occurs as a free element in nature. It is much too active. It always occurs as part of a compound. The most common source of sodium in the Earth is halite. Halite is nearly pure sodium chloride (NaCl). It is also called rock salt.

Halite can be found in underground deposits similar to coal mines. Those deposits were formed when ancient oceans evaporated (dried up), leaving sodium chloride behind. Earth movements eventually buried those deposits. Now they can be mined to remove the sodium chloride.

Sodium chloride can also be obtained from seawater and brine. Brine is similar to seawater, but it contains more dissolved salt. Removing sodium chloride from seawater or brine is easy.

Sodium stored in oil to prevent its reaction with the surrounding air.

All that is needed is to let the water evaporate. The sodium chloride is left behind. It only needs to be separated from other chemicals that were also dissolved in the water.

Isotopes

There is only one naturally occurring isotope of sodium, sodium-23. Isotopes are two or more forms of an element. Isotopes differ from each other according to their mass number. The number written to the right of the element's name is the mass number. The mass number represents the number of protons plus neutrons in the nucleus of an atom of the element. The number of protons determines the element, but the number of neutrons in the atom of any one element can vary. Each variation is an isotope.

Six radioactive isotopes of sodium are known also. A radioactive isotope is one that breaks apart and gives off some form of radiation. Radioactive isotopes are produced when very small particles are fired at atoms. These particles stick in the atoms and make them radioactive.

Two radioactive isotopes of sodium—sodium-22 and sodium-24—are used in medicine and other applications. They can be used as tracers to follow sodium in a person's body. A tracer is a radioactive isotope whose presence in a system can easily be detected. The isotope is injected into the system at some point. Inside the system, the isotope gives off radiation. That radiation can be followed by means of detectors placed around the system.

Sodium-24 also has non-medical applications. For example, it is used to test for leaks in oil pipe lines. These pipe lines are usually buried underground. It may be difficult to tell when a pipe begins to leak. One way to locate a leak is to add some sodium-24 to the oil. If oil leaks out of the pipe, so does the sodium-24. The leaking oil may not be visible, but the leaking sodium-24 is easily detected. It is located by instruments that are designed to detect radiation.

Extraction

One way to obtain pure sodium metal is by passing an electric current through molten (melted) sodium chloride:

$$2NaCl \xrightarrow{\text{electric current}} 2Na + Cl_2$$

This method is similar to the one used by Humphry Davy in 1808.

But there is not much demand for sodium metal. Sodium compounds are much more common. A second and similar method is used to make a compound known as sodium hydroxide (NaOH). The sodium hydroxide is then used as a starting point for making other sodium compounds.

The method for making sodium hydroxide is called the chlor-alkali process. The name comes from the fact that both chlorine and an alkali metal (sodium) are produced at the same time. In this case, an electric current is passed through a solution of sodium chloride dissolved in water:

$$2NaCl + 2H_2O \xrightarrow{\text{electric current}} Cl_2 + H_2 + 2NaOH$$

Three useful products are obtained from this reaction: chlorine gas (Cl_2), hydrogen gas (H_2), and sodium hydroxide (NaOH). The chlor-alkali process is one of the most important industrial processes used today.

Uses

Sodium metal has a relatively small, but important, number of uses. For example, it is sometimes used as a heat exchange medium in nuclear power plants. A heat exchange medium is a material that picks up heat in one place and carries it to another place. Water is a common heat exchange medium. Some home furnaces burn oil or gas to heat water that travels through pipes and radiators in the house. The water gives off its heat through the radiators.

Sodium does a similar job in nuclear power plants. Heat is produced by nuclear fission reactions at the core (center) of a nuclear reactor. In a nuclear fission reaction, large atoms break down to form smaller atoms. As they do so, large amounts of heat energy are given off.

Liquid sodium is sealed into pipes that surround the core of the reactor. As heat is generated, it is absorbed (taken up) by the sodium. The sodium is then forced through the pipes into a nearby room. In that room, the sodium pipes are wrapped around pipes filled with water. The heat in the sodium converts the water to steam. The steam is used to operate devices that generate electricity.

Another use of sodium metal is in producing other metals. For example, sodium can be combined with **titanium** tetrachloride ($TiCl_4$) to make titanium metal:

$$4Na + TiCl_4 \rightarrow 4NaCl + Ti$$

Sodium is also used to make artificial rubber. (Real rubber is made from the collected sap of rubber trees and is expensive.) The starting material for artificial rubber is usually a small molecule. The small molecule reacts with itself over and over again. It becomes a much larger molecule called a polymer. The polymer is the material that makes up the artificial rubber. Sodium metal is used as a catalyst in this reaction. A catalyst is a substance used to speed up or slow down a chemical reaction without undergoing any change itself.

Sodium is frequently used in making light bulbs. Sodium is first converted to a vapor (gas) and injected into a glass bulb. An electric current is passed through a wire or filament in the gas-filled bulb. The electric current causes the sodium vapor to

The combination of an electric current and sodium vapor produces a yellowish glow in street lamps.

give off a yellowish glow. Many street lamps today are sodium vapor lamps. Their advantage is that they do not produce as much glare as do ordinary lights.

Compounds

Almost all sodium compounds dissolve in water. When it rains, sodium compounds dissolve and are carried into the ground. Eventually, the compounds flow into rivers and then into the oceans. The ocean is salty partly because sodium compounds have been dissolved for many centuries.

But that means that finding sodium compounds on land is somewhat unusual. They tend to be more common in desert areas because deserts experience low rainfall. So sodium compounds are less likely to be washed away. Huge beds of salt and sodium carbonate are sometimes found in desert areas.

Dozens of sodium compounds are used today in all fields. Some of the most important of these compounds are discussed below.

Almost all sodium compounds dissolve in water. They tend to be more common in desert areas because deserts experience low rainfall.

Sodium chloride (NaCl). The most familiar use of sodium chloride is as a flavor enhancer in food. It is best known as table salt. Large amounts of sodium chloride are also added to prepared foods, such as canned, bottled, frozen, and dried foods. One purpose of adding sodium chloride to these foods is to improve their flavors. But another purpose is to prevent them from decaying. Sodium chloride kills bacteria in foods. It has been used for hundreds of years as a food preservative. The "pickling" or "salting" of a food, for example, means the adding of salt to that food to keep it from spoiling.

This process is one reason people eat so much salt in their foods today. Most people eat a lot of prepared foods. Those prepared foods contain a lot of salt. People are often not aware of all the salt they take in when they eat such foods.

Sodium chloride is also the starting point for making other sodium compounds. In fact, this application is probably the number one use for sodium chloride.

Sodium carbonate (Na_2CO_3). Sodium carbonate is also known by other names, such as soda, soda ash, sal soda, and washing soda. It is also used as the starting point in making other sodium compounds. A growing use is in water purifica-

Sodium chloride (table salt) crystals.

tion and sewage treatment systems. The sodium carbonate is mixed with other chemicals that react to form a thick, gooey solid. The solid sinks to the bottom of a tank, carrying impurities present in water or waste water.

Sodium carbonate is also used to make a very large number of commercial products, such as glass, pulp and paper, soaps and detergents, and textiles.

Sodium bicarbonate (NaHCO3). When sodium bicarbonate is dissolved in water, it produces a fizzing reaction. That reaction can be used in many household situations. For example, the fizzy gas can help bread batter rise. The "rising" of the batter is caused by bubbles released when sodium bicarbonate (baking soda) is added to milk in the batter. Certain kinds of medications, such as Alka-Seltzer, also include sodium bicarbonate. The fizzing is one of the effects of taking Alka-Seltzer that helps settle the stomach. Sodium bicarbonate is also used in mouthwashes, cleaning solutions, wool and silk cleaning systems, fire extinguishers, and mold preventatives in the timber industry.

Examples of lesser known compounds are as follows:

sodium alginate ($NaC_6H_7O_6$): a thickening agent in ice cream and other prepared foods; manufacture of cement; coatings for paper products; water-based paints

sodium bifluoride (KHF_2): preservative for animal specimens; antiseptic (germ-killer); etching of glass; manufacture of tin plate

sodium diuranate, or "**uranium** yellow" ($Na_2U_2O_7$): used to produce yellowish-orange glazes for ceramics

sodium fluorosilicate (Na_2SiF_6): used to make "fluoride" toothpastes that protect against cavities; insecticides and rodenticides (rat-killers); moth repellent; wood and leather preservative; manufacture of laundry soaps and "pearl-like" enamels

sodium metaborate ($NaBO_2$): herbicide

sodium paraperiodate ($Na_3H_2IO_6$): helps tobacco to burn more completely and cleanly; helps paper products retain strength when wet

sodium stearate ($NaOOCC_{17}H_{35}$): keeps plastics from breaking down; waterproofing agent; additive in toothpastes and cosmetics

sodium **zirconium** glycolate ($NaZrH_3(H_2COCOO)_3$): deodorant; germicide (germ-killer); fire-retardant

A common compound of sodium, sodium bicarbonate, produces a fizzing reaction. It is an ingredient in such medications as Alka-Seltzer.

Dietary concerns

People sometimes talk about the amount of "sodium" in their diet. Or they may refer to the amount of "salt" in their diet. The two terms are similar, but not exactly alike. In the body, sodium occurs most often as sodium chloride. A common name for sodium chloride is salt.

The Committee on Dietary Allowance of the U.S. Food and Nutrition Board recommends that a person take in about 1,100 to 3,300 milligrams of sodium per day. The human body actually needs only about 500 milligrams of sodium. Studies show that the average American takes in about 2,300 to 6,900 milligrams of sodium per day.

This high level of sodium intake troubles many health experts. Too much sodium can affect the body's ability to digest fats, for example. The most serious problem, however, may be hypertension. Hypertension is another name for "high blood pressure." A person with high blood pressure may be at risk for stroke, heart attack, or other serious health problems.

Health effects

Sodium has a number of important functions in plants, humans, and animals. In humans, for example, sodium is involved in controlling the amount of fluid present in cells. An excess or lack of sodium can cause cells to gain or lose water. Either of these changes can prevent cells from carrying out their normal functions.

Sodium is also involved in sending nerve messages to and from cells. These impulses control the way muscles move. Again, an excess or lack of sodium can result in abnormal nerve and muscle behavior. Sodium is also needed to control the digestion of foods in the stomach and intestines.

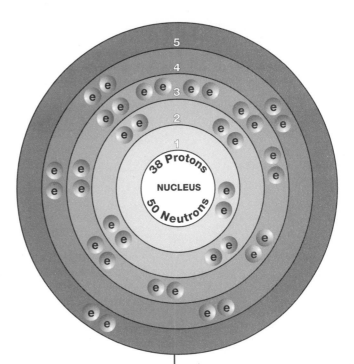

STRONTIUM

Overview

Strontium is a member of the alkaline earth metals. The alkaline earth metals make up Group 2 (IIA) of the periodic table. The periodic table is a chart that shows how chemical elements are related to one another. Other alkaline metals include **beryllium, magnesium, calcium, barium,** and **radium.** Strontium occupies a middle position in the family. Chemically, it is more active than calcium or magnesium, above it in the periodic table. But it is less active than barium, below it in Group 2.

The existence of strontium was first recognized in 1790 by Irish physician Adair Crawford (1748–95). However, the element was not prepared in pure form until nearly 20 years later by English chemist Humphry Davy (1778–1829). (See sidebar on Davy in the **calcium** entry in Volume 1.)

By far the major use of strontium is in the production of color television tubes. It is also used in the manufacture of ceramics and specialty glass. One of its radioactive isotopes is used in industry and medical studies.

Discovery and naming

Adair Crawford was trained as a physician. However, he was also interested in chemical research. For a period of time, he

SYMBOL
Sr

ATOMIC NUMBER
38

ATOMIC MASS
87.62

FAMILY
Group 2 (IIA)
Alkaline earth metal

PRONUNCIATION
STRONT-she-um

was on the staff at St. Thomas's Hospital in London, England, and a professor of chemistry at Woolwich University.

In 1790, he began studing certain minerals that were on display at St. Thomas's. These minerals were thought to be a form of baryte. Baryte is a mineral from which the element barium is obtained.

But Crawford found that some of the minerals did not behave as he expected. They did not have the properties of barium minerals. He concluded that the minerals contained a new element. He called the element strontia. He named it after a **lead** mine in Strontia, Scotland, from which the samples came.

Strontia was later found to be a compound of strontium and **oxygen.** In 1808, Davy found a way to produce pure strontium metal. He passed an electric current through molten (melted) strontium chloride. The electric current broke the compound into its two elements:

$$SrCl_2 \xrightarrow{\text{electric current}} Sr + Cl_2$$

Physical properties

Strontium is a silvery-white, shiny metal. When exposed to air, it combines with oxygen to form a thin film of strontium oxide (SrO). The film gives the metal a yellowish color.

Strontium has a melting point of about 757°C (1,395°F) and a boiling point of 1,366°C (2,491°F). Its density is 2.6 grams per cubic centimeter.

Chemical properties

Strontium is so active it must be stored under kerosene or mineral oil. In this way, the metal does not come into contact with air. In a finely divided or powdered form, strontium catches fire spontaneously and burns vigorously. Strontium is active enough to combine even with **hydrogen** and **nitrogen** when heated. The compounds formed are strontium hydride (SrH_2) and strontium nitride (Sr_3N_2). Strontium also reacts with cold water and with acids to release hydrogen gas:

$$Sr + 2H_2O \rightarrow Sr(OH)_2 + H_2$$

WORDS TO KNOW

Alkaline earth metal an element found in Group 2 (IIA) of the periodic table

Alloy a mixture of two or more metals with properties different from those of the individual metals

Isotopes two or more forms of an element that differ from each other according to their mass number

Periodic table a chart that shows how chemical elements are related to each other

Radioactive isotope an isotope that breaks apart and gives off some form of radiation

Occurrence in nature

Strontium is a relatively abundant element in the Earth's crust. It ranks about 15th among the elements found in the Earth. That makes it about as abundant as **fluorine** and its alkaline earth partner, barium.

The most common minerals containing strontium are celestine and strontianite. Celestine contains primarily strontium sulfate ($SrSO_4$), while strontianite contains mostly strontium carbonate ($SrCO_3$). Important world sources of strontium are Mexico, Spain, Turkey, and Iran. A small amount of strontium is also obtained from mines in California and Texas.

Isotopes

Four isotopes of strontium occur in nature. They are strontium-84, strontium-86, strontium-87, and strontium-88. Isotopes are two or more forms of an element. Isotopes differ from each other according to their mass number. The number written to the right of the element's name is the mass number. The mass number represents the number of protons plus neutrons in the nucleus of an atom of the element. The number of protons determines the element, but the number of neutrons in the atom of any one element can vary. Each variation is an isotope.

About ten radioactive isotopes of strontium are known also. A radioactive isotope is one that breaks apart and gives off some form of radiation. Radioactive isotopes are produced when very small particles are fired at atoms. These particles stick in the atoms and make them radioactive.

One radioactive isotope of strontium, strontium-90, is of special interest. It is a toxic substance which, at one time, was the cause of great concern because of its connection to atomic bomb testing. (See accompanying sidebar.)

Today, strontium-90 has a number of useful applications. For example, it is used to monitor the thickness of materials. Metal sheeting for construction must be the same thickness throughout. The sheeting is carried along on a conveyor belt beneath a small container of strontium-90. The isotope gives off radiation, some of which passes through the metal sheeting. The thicker the sheeting, the less radiation gets through. The thinner the sheeting, the more radiation gets through. A radiation detector is placed below the conveyor belt. The detector mea-

Poison from the sky: strontium-90

Strontium-90 is a radioactive isotope produced during the explosion of atomic weapons, such as an atomic bomb. In the 1950s and 1960s, the United States, the then-Soviet Union, China, and a few other nations tested atomic bombs in the atmosphere. Whenever one of these bombs exploded, some strontium-90 was thrown high into the atmosphere. After a short time, the strontium-90 settled to the ground where it was absorbed by growing plants. When cattle, sheep, and other domestic animals ate the plants, they also took strontium-90 into their bodies.

Strontium is just below calcium on the periodic table. That means that strontium behaves in much the same way that calcium does. Calcium eaten by humans and animals goes primarily to building bones and teeth. TV advertisements frequently recommend that young children drink milk. That's because milk contains calcium. It is used to build bones and teeth in growing children.

So any strontium that enters an animal's body is also used to build bones and teeth. The bad news is that strontium-90 is radioactive. It gives off radiation that kills or damages living cells. It can also cause those cells to begin growing out of control. Out-of-control cells lead to cancer. Strontium-90 in bones and teeth is a built-in time bomb. As long as it remains in the body, it has the potential for causing cancer in people and animals.

The threat posed by strontium-90 is one reason that nations agreed to begin testing nuclear weapons underground. It also helped world leaders realize that they needed to stop the testing of nuclear weapons entirely. It led to some degree to the agreements signed in the 1980s among the United States, Soviet Union, and other nations to give up atomic bomb testing entirely.

sures the amount of radiation passing through the sheeting. An inspector monitors the reading and makes adjustments to the manufacturing equipment to maintain the right thickness.

Strontium-90 is used for a number of other industrial applications, all based on the same principle. For instance, strontium-90 is used to measure the density of silk and tobacco products.

Strontium-90 has medical applications. A recent advance is to use the isotope for the control of pain. People who have cancer of the bone often experience terrible pain. At one time, the only treatment was medication. But those drugs often had unpleasant side-effects, such as nausea, dizziness, or depression.

Injecting strontium-90 into a person's body is now an alternative to the use of drugs. The strontium-90 deposits in the

bones, just as the calcium does. Within bones, the isotope stops pain signals being sent to the brain.

There are other medical applications for radioactive strontium isotopes. Strontium-90 is used to treat a variety of eye disorders. And strontium-85 and strontium-87m are used to study the condition of bones in a person's body.

Extraction
Most strontium metal is still obtained by the method used by Davy. An electric current is passed through molten (melted) strontium chloride.

Uses and compounds
Strontium and its compounds have relatively few commercial uses. The pure metal is sometimes combined with other metals to form alloys. An alloy is made by melting and mixing two or more metals. The mixture has different properties than the individual metals. Compounds of strontium are sometimes used to color glass and ceramics. They give a beautiful red color to these materials. Strontium compounds also provide the brilliant red color of certain kinds of fireworks.

Health effects

Most strontium compounds are regarded as harmless to plants and animals. A few, such as strontium chloride ($SrCl_2$) and strontium iodide (SrI_2), are somewhat toxic.

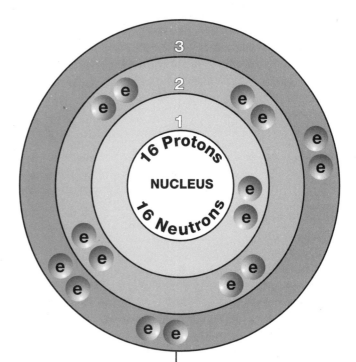

SULFUR

Overview

Sulfur belongs to the chalcogen family. Other members of the family are **oxygen, selenium, tellurium,** and **polonium.** These elements make up Group 16 (VIA) of the periodic table. The periodic table is a chart that shows how chemical elements are related to each other.

The term chalcogen comes from two Greek words meaning "ore forming." An ore is a naturally occurring mineral used as a source for an element. Many ores are compounds of a metal and oxygen or a metal and sulfur. Compounds that contain two elements, one of which is sulfur, are called sulfides. For example, a beautiful gold-colored mineral is called pyrite, or "fool's gold," because it looks so much like real gold. Pyrite is iron sulfide (FeS_2).

Sulfur was known to ancient peoples. Its physical and chemical properties are very distinctive. It often occurs as a brilliant yellow powder. When it burns, it produces a clear blue flame and a very strong odor.

Sulfur, also spelled as sulphur, is a very important element in today's world. Its most important use is in the manufacture of

SYMBOL
S

ATOMIC NUMBER
16

ATOMIC MASS
32.064

FAMILY
Group 16 (VIA)
Chalcogen

PRONUNCIATION
SUL-fur

sulfuric acid (H_2SO_4). There is more sulfuric acid made than any other chemical in the world. It has an enormous number of important uses.

Discovery and naming

Sulfur must have been well known to ancient peoples. They sometimes referred to it as brimstone. Sulfur sometimes occurs in bright yellow layers on the top of the earth. It has a sharp, offensive odor. When it burns, it gives off a strong, suffocating smell. The odor is like that produced when a match is struck.

The Bible mentions brimstone in a number of places. For example, Sodom and Gomorrah were two towns destroyed by God for the wicked ways of their citizens: "The Lord rained upon Sodom and upon Gomorrah brimstone and fire."

But ancient people certainly did not think about sulfur the way modern chemists do. In fact, they used the word "element" to talk about anything that was basic. Ancient Greek philosophers, for example, thought that everything consisted of four elements: earth, fire, water, and air. Other philosophers thought there were only two elements: sulfur and **mercury.**

But early thinkers were often confused as to what they meant by the word "sulfur." They often were talking about anything that burned and gave off large amounts of smoke. To them, "sulfur" was really a "burning substance." It took centuries for scientists to identify sulfur as an element.

Physical properties

Sulfur exists in two allotropic forms. Allotropes are forms of an element with different physical and chemical properties. The two forms of sulfur are known as α-form and β-form (the Greek letters alpha and beta, respectively). Both allotropes are yellow, with the α-form a brighter yellow and the β-form a paler, whitish-yellow. The α-form changes to the β-form at about 94.5°C (202°F). The α-form can be melted at 112.8°C (235.0°F) if it is heated quickly. The β-form has a melting point of 119°C (246°F). The boiling point of the α-form is 444.6°C (832.3°F).

The two allotropes have densities of 2.06 grams per cubic centimeter (α-form) and 1.96 grams per cubic centimeter (β-form). Neither allotrope will dissolve in water. Both are soluble

in other liquids, such as benzene (C_6H_6), **carbon** tetrachloride (CCl_4), and carbon disulfide (CS_2).

Solid sulfur.

Another allotrope of sulfur is formed when the element is melted. This allotrope has no crystalline shape. It looks like a dark brown, thick, melted plastic.

Chemical properties

Sulfur's most prominent chemical property is that it burns. When it does so, it gives off a pale blue flame and sulfur dioxide (SO_2) gas. Sulfur dioxide has a very obvious strong, choking odor.

Sulfur also combines with most other elements. Sometimes it combines with them easily at room temperature. In other cases, it must be heated. The reaction between **magnesium** and sulfur is typical. When the two elements are heated, they combine to form magnesium sulfide (MgS):

$$Mg + S \xrightarrow{\text{heated}} MgS$$

Sulfur sometimes occurs in bright yellow layers on the top of the earth. It has a sharp, offensive odor.

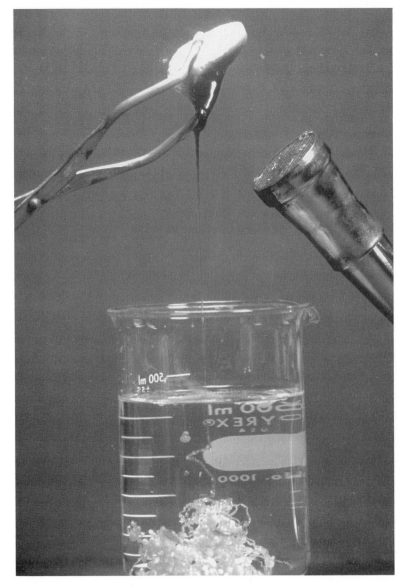

A chemical reaction involving sulfur.

Sulfur also combines with **hydrogen** gas:

$$H_2 + S \xrightarrow{\text{heated}} H_2S$$

The compound formed in this reaction is hydrogen sulfide (H_2S). Hydrogen sulfide has one of the best known odors of all compounds. It smells like rotten eggs. Hydrogen sulfide is added to natural gas (methane) used in homes for cooking and heating. Methane is odorless. So the unique smell of hydrogen sulfide makes it easy to know when there is a methane leak.

Occurrence in nature

At one time, sulfur occurred in layers along the Earth's surface. They were easy for humans to find and take. Deposits like these are more difficult to find today. One place they still occur is in the vicinity of volcanoes. Sulfur is released from volcanoes as a gas. When it reaches the cold air, it changes back to a solid. It forms beautiful yellow deposits along the edge of a volcano.

Large supplies of sulfur still occur underground. They are removed by the Frasch process (see accompanying sidebar).

Sulfur also occurs in a number of important minerals. Some examples are barite, or barium sulfate ($BaSO_4$); celestite, or **strontium** sulfate ($SrSO_4$); cinnabar, or mercury sulfide (HgS); galena, or **lead** sulfide (PbS); pyrites, or **iron** sulfide (FeS_2); sphalerite, or **zinc** sulfide (ZnS); and stibnite, or **antimony** sulfide (Sb_2S_3).

Sulfur occurs in the vicinity of volcanoes.

The abundance of sulfur in the Earth's crust is thought to be about 0.05 percent. It ranks about number 16 among the elements in terms of their abundance in the earth. It is more abundant than carbon, but less abundant than barium or strontium.

The largest producers of sulfur in the world are the United States, Canada, China, Russia, Mexico, and Japan. In 1996, the United States produced about 11,800,000 metric tons of sulfur. It is mined in 30 states, Puerto Rico, and the U.S. Virgin Islands.

Isotopes

There are four naturally occurring isotopes of sulfur: sulfur-32, sulfur-33, sulfur-34, and sulfur-36. Isotopes are two or more forms of an element. Isotopes differ from each other according to their mass number. The number written to the right of the element's name is the mass number. The mass number represents the number of protons plus neutrons in the nucleus of an atom of the element. The number of protons determines the element, but the number of neutrons in the atom of any one element can vary. Each variation is an isotope.

Six radioactive isotopes of sulfur are known also. A radioactive isotope is one that breaks apart and gives off some form of radiation. Radioactive isotopes are produced when very small

The Frasch method of removing sulfur

The Frasch method is one of the most famous mining systems ever invented. It was developed by German-American chemist Herman Frasch (1851–1914) in 1887.

The Frasch method is based on the low melting point of sulfur. The element melts at a temperature slightly higher than that of boiling water (100°C). Here is how the method works:

A set of three nested pipes (one inside each other) is sunk into the ground. The innermost pipe has a diameter of about an inch. The middle pipe has a diameter of about four inches. And the outer pipe has a diameter of about eight inches.

A stream of superheated water is injected into the outer pipe. Superheated water is water that is hotter than its boiling point, but that has not started to boil. Superheated water can be made by raising the pressure on the water. Its temperature can reach 160°C (320°F).

The superheated water passes down the outer pipe into the underground sulfur, causing it to melt. The molten (melted) sulfur forms a lake at the bottom of the pipe.

At the same time, a stream of hot air under pressure is forced down the innermost (one-inch) pipe. The hot air stirs up the molten sulfur and hot water at the bottom of the pipe. A foamy, soupy mixture of sulfur and water is formed. The mixture is forced upward through the middle pipe. When it reaches the surface, it is collected. The sulfur cools and separates from the water.

particles are fired at atoms. These particles stick in the atoms and make them radioactive.

One radioactive isotope of sulfur, sulfur-35, is used commercially. In medicine, the isotope is used to study the way fluids occur inside the body. It also has applications in research as a tracer. A tracer is a radioactive isotope whose presence in a system can easily be detected. The isotope is injected into the system at some point. Inside the system, the isotope gives off radiation. That radiation can be followed by means of detectors placed around the system.

As an example, a company that makes rubber tires might want to know what happens to the sulfur added to tires. Sulfur-35 is added to rubber along with non-radioactive sulfur. Researchers follow the radioactive isotope in the tires to see what happens to the sulfur when the tires are used.

Similar applications of sulfur-35 involve studying sulfur in steel when it is made, seeing how sulfur affects the way engines operate, following what happens when proteins (which contain sulfur) are digested, and learning how drugs that contain sulfur are processed in the body.

Extraction

Like coal, sulfur sometimes occurs in thick layers under ground. One way to remove sulfur would be to mine it the way coal is mined. But a much easier method for removing sulfur from the ground is the Frasch method (see accompanying sidebar).

Uses

Sulfur has relatively few uses as an element. One of the most important of those uses is in vulcanization. Vulcanization is the process of adding sulfur to rubber to make it stiff and hard. It keeps the rubber from melting as it gets warmer. The discovery of vulcanization by Charles Goodyear (1800–60) in 1839 is one of the greatest industrial accomplishments of modern times.

Some powdered sulfur is also used as an insecticide. It can be spread on plants to kill or drive away insects that feed on the plants. By far the majority of sulfur is used, however, to make sulfur compounds. The most important of these is sulfuric acid (H_2SO_4).

Compounds

Nearly 90 percent of all sulfur produced goes into sulfuric acid. Sulfuric acid is the number one chemical in the world in terms of the amount produced. Each year, almost twice as much sulfuric acid is made as the next highest chemical, **nitrogen.** In 1996, more than 45 million tons of sulfuric acid were produced in the United States alone.

The greatest portion, nearly 75 percent, of sulfuric acid is used to make fertilizers. The next most important use, 10 percent, is in the petroleum industry. Other important uses of sulfuric acid are in the treatment of **copper** ores; the production of paper and paper products; the manufacture of other agricultural chemicals; and the production of plastics, synthetic rubber, and other synthetic materials.

Sulfuric acid is also used in smaller amounts to make explosives, water treatment chemicals, storage batteries, pesticides,

Vulcanization is the process of adding sulfur to rubber to make it stiff and hard.

Sulfur

Reaction of sulfuric acid and sugar.

drugs, synthetic fibers, and many other chemicals used in everyday life.

Health effects

The cleansing power of sulfur has been known for many centuries. At one time, ancient physicians burned sulfur in a house to cleanse it of impurities. Creams made with sulfur were used to treat infections and diseases. In fact, sulfur is still used to treat certain medical problems. Sulfur is prepared in one of three forms. Precipitated sulfur (milk of sulfur) is made by boiling sulfur with lime. Sublimed sulfur (flowers of sulfur) is pure sulfur powder. And washed sulfur is sulfur treated with ammonia water. Washed sulfur is used to kill parasites (organisms that live on other organisms) such as fleas and ticks. It is also used as a laxative, a substance that helps loosen the bowels.

Sulfur is a macronutrient for both plants and animals. A macronutrient is an element needed in relatively large amounts to insure the good health of an organism. Sulfur is used to make proteins and nucleic acids, such as DNA. It also occurs in many essential enzymes. Enzymes are chemicals that make chemical reactions occur more quickly in cells. Humans usually have no problem getting enough sulfur in their diets. Eggs and meats are especially rich in sulfur.

A person who does not get enough sulfur in his or her diet develops certain health problems. These include itchy and flaking skin and improper development of hair and nails. Under very unusual conditions, a lack of sulfur can lead to death. Such conditions would be very rare, however.

Plants require sulfur for normal growth and development. When plants do not get enough sulfur from the soil, their young leaves start to turn yellow. Eventually, this yellowing extends to the whole plant. The plant may develop other diseases as a result.

The cleansing power of sulfur has been known for many centuries.

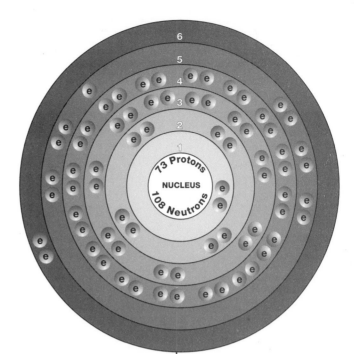

6
5
4
3
2
1

73 Protons

NUCLEUS

108 Neutrons

TANTALUM

Overview

Tantalum is a transition metal in Group 5 (VB) of the periodic table. The periodic table is a chart that shows how chemical elements are related to one another. Tantalum is one of the most inert metals known. An inert material is one that does not react with most other chemicals. Most metals, for example, dissolve in acids, but tantalum is not affected by acids or other strong chemicals. For this reason, tantalum is used to make chemical, medical, and dental equipment.

Credit for the discovery of tantalum goes to Swedish chemist and mineralogist Anders Gustaf Ekeberg (1767–1813). Ekeberg announced his discovery in 1802. However, chemists were uncertain about Ekeberg's new element for many years. They believed that another element, **niobium,** might be present along with tantalum. In fact, it was not until 50 years later that chemists could be sure that tantalum and niobium were really two different elements.

Discovery and naming

In 1801, English chemist Charles Hatchett (1765–1847) discovered a new element that he named niobium. A year later, Ekeberg discovered a new element that he named tantalum.

SYMBOL
Ta

ATOMIC NUMBER
73

ATOMIC MASS
180.9479

FAMILY
Group 5 (VB)
Transition metal

PRONUNCIATION
TAN-tuh-lum

The two names are related. Niobium was named for the mythical daughter of Tantalus, Niobe.

Tantalus was a son of Zeus, the major Greek god. Zeus decided to punish his son for giving the gods' secrets to humans. He forced Tantalus to stand in a vat filled with water up to his chin. Whenever Tantalus bent to take a drink, the water dropped a little lower so he could never get his drink. Ekeberg said that his new element was like Tantalus. When placed in acid, it did not take up (react with) the acid.

Most chemists thought that the two men's discoveries were one and the same. The two elements reacted exactly like each other. They could not see how tantalum was different from niobium. For more than 40 years, the general belief was that Ekeberg and Hatchett had discovered the same element.

In 1844, however, German chemist Heinrich Rose (1795–1864) announced new evidence. He found that tantalic acid (H_3TaO_4) made from tantalum and niobic acid (H_3NbO_4) made from niobium were definitely different from each other. He confirmed that Ekeberg and Hatchett had really discovered two different elements.

Physical properties

Tantalum is a very hard, malleable, ductile metal. Malleable means capable of being hammered into thin sheets. Ductile means capable of being drawn into thin wires. The metal has a silvery-bluish color when unpolished, but a bright silvery color when polished. It has a melting point of 2,996°C (5,425°F) and a boiling point of 5,429°C (9,804°F). It has the third highest melting point of all elements, after **tungsten** and **rhenium.** Tantalum's density is 16.69 grams per cubic centimeter.

Chemical properties

Tantalum is one of the most unreactive metals. At room temperature, it reacts only with **fluorine** gas and certain fluorine compounds. Fluorine, a non-metal, is the most active element. At higher temperatures, tantalum becomes more active. Above about 150°C (300°F), it reacts with acids and alkalis. An alkali is the chemical opposite of an acid.

Occurrence in nature

Tantalum ranks about number 50 among elements found in the Earth's crust. It is slightly more common than tungsten, but less common than **arsenic.** Its abundance is probably about 1.7 parts per million in the earth. The element is most commonly found in the minerals columbite, tantalite, and microlite. It always occurs with niobium.

The only source of tantalum in North America is a mine located at Bernic Lake in the province of Manitoba in Canada. Most of the tantalum used in the United States comes from Australia, Germany, Thailand, and Brazil.

Isotopes

There are two naturally occurring isotope of tantalum, tantalum-180 and tantalum-181. Isotopes are two or more forms of an element. Isotopes differ from each other according to their mass number. The number written to the right of the element's name is the mass number. The mass number represents the number of protons plus neutrons in the nucleus of an atom of the element. The number of protons determines the element, but the number of neutrons in the atom of any one element can vary. Each variation is an isotope.

Tantalum-181 is radioactive. A radioactive isotope is one that breaks apart and gives off some form of radiation. The half life of a radioactive element is the time it takes for half of a sample of the element to break down. Tantalum-181 has a half life of more than one trillion years. It makes up about 0.01 percent of all natural tantalum.

More than a dozen radioactive isotopes of tantalum have been made artificially. None of these isotopes has any commercial application.

Extraction

After tantalum ores are taken from the earth, they are converted to tantalum **potassium** fluoride (K_2TaF_7). Pure tantalum is then obtained from this compound by passing an electric current through it.

Uses

The primary use of tantalum metal is in making capacitors. A capacitor is an electrical device similar to a battery. It can be

Tantalum is one of the most unreactive metals. At room temperature, it reacts only with fluorine gas and certain fluorine compounds.

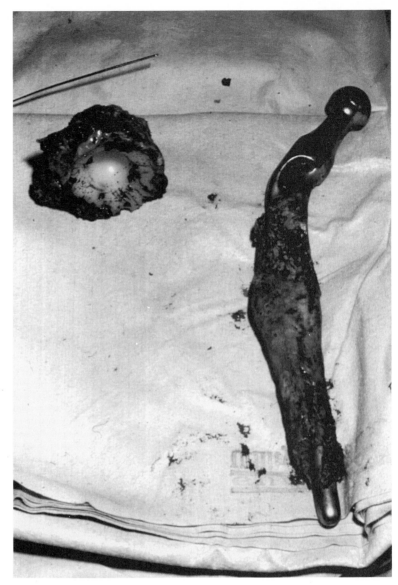

Tantalum alloys are used to make artificial joints, such as an artificial hip shaft (left) and socket.

given an electrical charge, which it then stores until needed. Capacitors are essential parts of nearly all electrical circuits. Semiconductor circuits, like those used in transistors, require tiny capacitors the size of grains of rice. Tantalum is one of the best metals for this purpose. Different kinds of capacitors are made for many different applications. They are used in military weapons systems, aircraft, space vehicles, communication systems, computers, and medical applications. For example, the smallest hearing aids are likely to have a tantalum capacitor.

Tantalum is also used in many different alloys. An alloy is made by melting and mixing two or more metals. The mixture has properties different from those of the individual metals. Tantalum alloys are used in laboratory equipment, weights for very precise balances, fountain and ball point pen points, and tools that have to operate at high speeds and temperatures.

Another application for tantalum alloys is in medical and dental applications. The metal has no effect on body tissues. It is used in artificial hips, knees, and other joints. Pins, screws, staples, and other devices used to holds bones together are also made of tantalum alloys.

Compounds

A few compounds of tantalum have some important uses. They are as follows:

tantalum carbide (TaC): a very hard material used for cutting tools and dies

tantalum disulfide (TaS_2): used in the form of a black powder, it acts as a solid lubricant, like powdered carbon

tantalum oxide (Ta_2O_5): used in the preparation of special types of glass; used in specialized lasers (devices for producing a very bright light of a single color)

Health effects

Tantalum and its compounds are not thought to pose a serious health hazards to humans and animals.

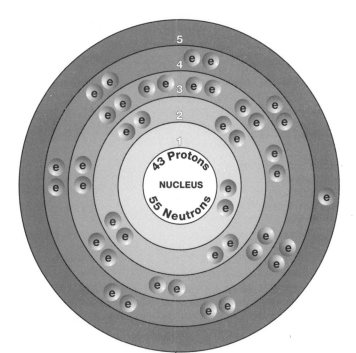

TECHNETIUM

Overview

Technetium is the only element below **uranium** that does not exist on Earth. It is a synthetic (man-made) element produced in a particle accelerator. For many years, chemists knew that an element with atomic number 43 must exist, based on an empty spot in the periodic table. The periodic table is a chart that shows how chemical elements are related to each other. By 1925, only five empty boxes remained. Five elements were still to be discovered.

A number of scientists tried to find those elements. In many cases, the results they announced were wrong. Such was the case with element 43. When the element's "discovery" was announced in 1925, chemists were excited, but no one else was able to repeat the discovery. An error had been made. In fact, it was not until more than ten years later that the element was finally produced. Then, it was created in a particle accelerator and not found on Earth.

Today, technetium has very few—but very important—uses. It is used in finding out more about diseases and health problems. It is also used to make steel stronger.

SYMBOL
Tc

ATOMIC NUMBER
43

ATOMIC MASS
97.9072

FAMILY
Group 7 (VIIB)
Transition metal

PRONUNCIATION
tek-NEE-she-um

Discovery and naming

In the 1920s, a team of researchers were looking for elements with atomic numbers 43 and 75. This team consisted of German chemists Walter Noddack (1893–1960), Ida Tacke (1896–1979), and Otto Berg. In 1925, the team announced that they had found both elements. They named element 43 masurium, after the region called Masurenland in eastern Germany, and element 75 **rhenium,** after the Rhineland, in western Germany. Although they were correct about rhenium, they were wrong about masurium. No other chemist was able to reproduce masurium.

So, chemists kept looking for element 43. It was finally discovered in the products of a particle accelerator experiment at the University of California at Berkeley. A particle accelerator is sometimes called an atom smasher. It accelerates small particles, such as protons, to very high speeds. The particles then collide with elements such as **gold, copper,** or **tin.** When struck by the particles, the targets often form new elements.

Element 43 was found by Italian physicist Emilio Segrè (1905–89) and his colleague Carlo Perrier. These researchers collected one ten-billionth of a gram of the element and studied some of its properties. They eventually gave the name technetium to the element, from the Greek word *technetos,* meaning "artificial." Technetium was the first element not found in the earth to be made artificially. It can now be made in much larger quantities, of at least a kilogram (two pounds) at a time.

Physical properties

Technetium is a silver-gray metal with a melting point of 2,200°C (4,000°F) and a density of 11.5 grams per cubic centimeter.

Chemical properties

Technetium is placed between **manganese** and **rhenium** on the periodic table. That would lead chemists to believe that its properties are like those of the other two elements. Experiments have shown this to be true. It reacts with some acids, but not others. It also reacts with **fluorine** gas and with **sulfur** at high temperatures.

Occurrence in nature

Some scientists believe that technetium will be found in very small amounts in the Earth's crust along with other radioactive materials, such as **uranium** and **radium.** However, it has never been found on Earth. It has, however, been found in certain types of stars. Its presence can be detected by analyzing the light produced by these stars.

Isotopes

All isotopes of technetium are radioactive. The most stable of these isotopes technetium-97, technetium-98, and technetium-99, have half lives of more than a million years. Isotopes are two or more forms of an element. Isotopes differ from each other according to their mass number. The number written to the right of the element's name is the mass number. The mass number represents the number of protons plus neutrons in the nucleus of an atom of the element. The number of protons determines the element, but the number of neutrons in the atom of any one element can vary. Each variation is an isotope.

The half life of a radioactive element is the time it takes for half of a sample of the element to break down. After about a million years, only 5 grams of a 10-gram sample of technetium-97 would remain. After another million years, only half of that, or 2.5 grams, would be left.

Technetium has never been found on Earth. It has, however, been found in certain types of stars.

Extraction

Technetium is now produced in nuclear reactors. Neutrons collide with atoms of uranium or **plutonium** to form new elements. The technetium is then converted to a compound called ammonium pertechnate (NH_4TcO_4). That compound is then treated with **hydrogen** gas to obtain pure technetium metal.

Uses

Technetium is used in steel alloys. An alloy is made by melting and mixing two or more metals. The mixture has properties different from those of the individual metals. Technetium-steel alloys are very resistant to corrosion or reaction with **oxygen** and other materials. No more than 50 parts per million of technetium to steel produces this property. Technetium-steel has limited uses, however, because technetium is radioactive. People cannot be exposed to technetium-steel directly.

Technetium as a medical tool

Technetium-99m is an almost ideal diagnostic tool. It has a half life of about six hours. After six hours, only half of it remains as technetium-99m. The rest has broken down into another element. After 24 hours, only one-sixteenth of the original isotope remains. It breaks down and disappears very quickly.

When injected into the body, it deposits in certain organs, such as the brain, liver, spleen, and kidney. It also deposits in the bones. Technetium-99m gives off radiation that can be detected very easily. The amount and location of the radiation indicates problems with an organ or bones. Technetium-99m sends out clear, easily observed signals for a short time. Then, it is eliminated from the body.

The only problem with technetium-99m is how to get it. If a doctor ordered technetium-99m from a supplier a great distance away, by the time the isotope arrived at the hospital, it would have almost completely broken down!

Instead, medical workers use a different isotope, molybdenum-99. When molybdenum-99 breaks down, it forms technetium-99m. But molybdenum-99 has a longer half life, almost three days.

When medical workers need technetium-99m, they bring in a container of molybdenum-99. They separate the technetium-99m as it is formed from the molybdenum-99. When molybdenum-99 is used in this way, it is called a "molybdenum cow."

Technetium is a popular diagnostic tool in medicine. The term diagnosis means to find out what is wrong with a person. (See accompanying sidebar.)

Compounds

There are no commercially important compounds of technetium.

Health effects

Since all forms of technetium are radioactive, the element must be handled with great care. Radiation damages or kills living cells. A significant exposure produces radiation sickness.

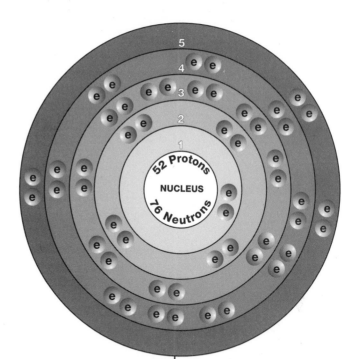

TELLURIUM

Overview

The elements that make up Group 16 (VIA) of the periodic table are sometimes called the *chalcogens*. This name comes from the Greek word for "bronze ore," *chalkos*. The first two elements in the family, **oxygen** and **sulfur,** are often found in such ores. Tellurium is the next to last member of that family. The periodic table is a chart that shows how chemical elements are related to one another.

The chalcogens are one of the most interesting families in the periodic table. The first member, oxygen, is a gas with very un-metal-like properties. The next two members of the family, sulfur and **selenium,** are solids, with increasingly metallic properties. Tellurium, near the bottom of the family, looks and behaves very much like most metals. The slow change of properties, from less metal-like to more metal-like, occurs in all families in the periodic table. But the change is seldom as dramatic as it is in the chalcogens.

Tellurium was discovered in 1782 by Austrian mineralogist Baron Franz Joseph Müller von Reichenstein (1740–1825 or 1826). The element seldom occurs in its pure state. It is usually found as a compound in ores of **gold, silver, copper, lead,**

SYMBOL
Te

ATOMIC NUMBER
52

ATOMIC MASS
127.60

FAMILY
Group 16 (VIA)
Chalcogen

PRONUNCIATION
tuh-LUHR-ee-um

mercury, or **bismuth.** The most common use of tellurium today is in specialized alloys. An alloy is made by melting and mixing two or more metals. The mixture has properties different from those of the individual metals. About three-quarters of all tellurium goes into alloys. The other two major uses of tellurium are in making chemicals and electrical equipment.

Discovery and naming

Müller discovered tellurium while studying gold taken from a mine in the Börzsöny Mountains of Hungary. He had received the gold from a colleague who thought that it contained an impurity. The colleague was unable to identify the impurity, but thought it might be "unripe gold."

The concept of "unripe gold" was invented before the birth of modern chemistry. Earlier scientists—called alchemists—thought that gold "grew" in the earth in much the same way that plants grow. They thought gold went through various stages, from lead to mercury to silver to gold. These metals were thought to be the same material in various stages of growth.

This view of tellurium is reflected in some of its older common names. It was also known as *aurum paradoxum* and as *metallum problematum*. The first name means "paradoxical gold," something that acts like gold, but really isn't. The second name means "the problem metal."

Müller held more modern views, however. He suspected that the impurity was not "unripe gold," but a new element. He conducted more than fifty tests on the new material over a three-year period. He came to have a clear understanding of the new element.

Many years later, Müller sent a sample of the new element to German chemist Martin Heinrich Klaproth (1743–1817). Klaproth confirmed Müller's discovery. He suggested the name tellurium, from the Latin word *tellus,* meaning "Earth."

Tellurium is often found with another element, selenium. That element was discovered 30 years later and named in honor of the Moon. In Latin, the moon is *selene.* The connection between tellurium and selenium is more clear now than it was when tellurium was first discovered.

WORDS TO KNOW

Alloy a mixture of two or more metals with properties different from those of the individual metals

Catalyst a substance used to speed up or slow down a chemical reaction without undergoing any change itself

Isotopes two or more forms of an element that differ from each other according to their mass number

Periodic table a chart that shows how chemical elements are related to each other

Radioactive isotope an isotope that breaks apart and gives off some form of radiation

Vulcanization the process by which soft rubber is converted to a harder, longer-lasting product

Physical properties

Tellurium is a grayish-white solid with a shiny surface. It has a melting point of 449.8°C (841.6°F) and a boiling point of 989.9°C (1,814°F). Its density is 6.24 grams per cubic centimeter. It is relatively soft. Although it has many metal-like properties, it breaks apart rather easily and does not conduct an electric current very well.

Chemical properties

Tellurium does not dissolve in water. But it does dissolve in most acids and some alkalis. An alkali is a chemical with properties opposite those of an acid. Sodium hydroxide (common lye, such as Drano) and limewater are examples of alkalis.

Tellurium also has the unusual property of combining with gold. Gold normally combines with very few elements. The compound formed between gold and tellurium is called gold telluride (Au_2Te_3). Much of the gold found in the earth occurs in the form of gold telluride.

Occurrence in nature

Tellurium is one of the rarest elements in the Earth's crust. Its abundance is estimated to be about 1 part per billion. That

places it about number 75 in abundance of the elements in the earth. It is less common than gold, silver, or **platinum.**

The most common mineral of tellurium is sylvanite. Sylvanite is a complex combination of gold, silver, and tellurium. Tellurium is obtained commercially today as a by-product in copper and lead refining.

Isotopes
Eight naturally occurring isotopes of tellurium are known. They are tellurium-120, tellurium-122, tellurium-123, tellurium-124, tellurium-125, tellurium-126, tellurium-128, tellurium-130. Isotopes are two or more forms of an element. Isotopes differ from each other according to their mass number. The number written to the right of the element's name is the mass number. The mass number represents the number of protons plus neutrons in the nucleus of an atom of the element. The number of protons determines the element, but the number of neutrons in the atom of any one element can vary. Each variation is an isotope.

At least a dozen radioactive isotopes of tellurium are known also. A radioactive isotope is one that breaks apart and gives off some form of radiation. Radioactive isotopes are produced when very small particles are fired at atoms. These particles stick in the atoms and make them radioactive.

None of the radioactive isotopes of tellurium have any commercial uses.

Extraction
A common method for obtaining tellurium is to pass an electric current through dissolved tellurium dioxide (TeO_2). The current breaks the tellurium dioxide down into oxygen and tellurium:

$$TeO_2 \xrightarrow{\text{electrical current}} Te + O_2$$

Uses and compounds
About 75 percent of all the tellurium produced today is used in alloys. Its most important alloy is a tellurium-steel alloy. It has better machinability than does steel without tellurium. Machinability means working with a metal: bending, cutting, shaping, turning, and finishing the metal, for example. Adding 0.04 percent tellurium to steel makes it much easier to work with.

> Tellurium has the unusual property of combining with gold. Gold normally combines with very few elements.

Tellurium is used in the manufacture of laser printers.

Tellurium is also added to copper to improve machinability. Tellurium-copper alloys are also easier to work with than pure copper. And the essential ability of copper to conduct an electric current is not affected. Tellurium is also added to lead. Tellurium-lead alloys are more resistant to vibration and fatigue than pure lead. Metal fatigue is the tendency of a metal to wear out and eventually break down after long use.

About 15 percent of all tellurium produced is used in the rubber and textile industries. It is important in the vulcanization of rubber, for example. Vulcanization is the process by which soft rubber is converted to a harder, longer-lasting product. Tellurium is also used as a catalyst in the manufacture of synthetic fibers. A catalyst is a substance used to speed up or slow down a chemical reaction without undergoing any change itself.

A growing application of tellurium is in a variety of electrical devices. For example, it is used to improve picture quality in photocopiers and printers. A compound of tellurium, **cadmium,** and mercury is also used in infrared detection systems. Infrared radiation is heat. It can be made visible with special glass. Some satellites orbiting the Earth study forests, crops, and other plant life by measuring the infrared radiation they give off.

Finally, a very small amount of tellurium is used for minor applications, such as a coloring agent in glass and ceramics and in blasting caps for construction projects.

Health effects
When taken internally, tellurium can have harmful effects. It may cause nausea, vomiting, and damage to the central nervous system. One interesting side effect is that it gives a garlicky-odor to the breath.

Tellurium gives a garlicky-odor to the breath.

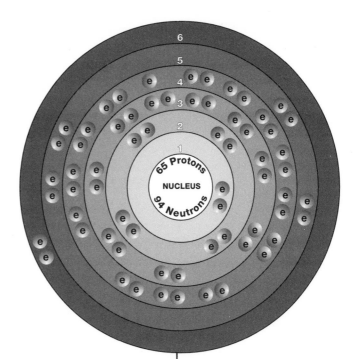

65 Protons
NUCLEUS
94 Neutrons

TERBIUM

Overview

Terbium is classified as a rare earth element. The term is misleading because terbium is more common than metals such as **silver** and **mercury.** The name "rare earth" meant something else to early chemists. It was used because the rare earth elements were very difficult to separate from each other. They were not "rare" in the Earth, but they were "rarely" used for anything.

Today, chemists have developed new ways of separating elements from each other. Terbium doesn't have a great many uses, but is easily available. One use is in television screens. It helps the screen display the colors more clearly.

Although the term "rare earth" is still used, the more proper name for terbium and its cousins is the lanthanides. The term lanthanide comes from the element **lanthanum** in Row 6 of the periodic table. The periodic table is a chart that shows how chemcial elements are related to one another.

Discovery and naming

Terbium was discovered during the great element hunt of the 1840s. That hunt began with a lucky discovery made in 1787.

SYMBOL
Tb

ATOMIC NUMBER
65

ATOMIC MASS
158.9254

FAMILY
Lanthanide
(rare earth metal)

PRONUNCIATION
TER-bee-um

A lieutenant in the Swedish army, Carl Axel Arrhenius (1757–1824), discovered an unusual black rock near the town of Ytterby, Sweden. Arrhenius passed the rock on to a friend of his, chemist Johan Gadolin (1760–1852). He analyzed the sample to see what elements it contained.

Gadolin first discovered an entirely new mineral that he named yttria, after the town of Ytterby. This discovery, however, was only the first in a long chain of puzzling new findings.

In 1843, Swedish chemist Carl Gustav Mosander (1797–1858) demonstrated that yttria was really a mixture of three other minerals. He called those minerals erbia, terbia, and yttria. All three of these names also came from the town of Ytterby. The ending -*a* on these names means they refer to minerals that occur in the earth. A mineral ending in -*a* usually refers to an element combined with **oxygen.** For example, soda is a combination of **sodium** and oxygen.

Mosander's research is long and complicated. Chemists did not have good equipment in the 1840s. They often made errors and were confused by their discoveries. For example, other chemists also studied the mineral yttria. When they did so, however, they got the names Mosander used mixed up. They called his terbia "erbia" and his erbia "terbia."

Mosander is given credit for discovering terbium even though he never saw the pure element. In 1886, French chemist Jean-Charles-Galissard de Marignac (1817–94) was the first to prepare pure terbium.

Physical properties

Terbium has the silver-gray luster typical of many metals. It is quite soft, however, and can be cut with a knife. It is also malleable and ductile, meaning it can be hammered into thin sheets and drawn into wires rather easily. The melting point of terbium is 1,356°C (2,473°F) and its boiling point is about 2,800°C (5,000°F). It has a density of 8.332 grams per cubic centimeter.

Chemical properties

Like many of its rare earth cousins, terbium is not very active. It does not react with oxygen in the air very easily. It does react with water slowly, however, and will dissolve in acids.

WORDS TO KNOW

Ductile capable of being drawn into thin wires

Fuel cell any system that uses chemical reactions to produce electricity

Isotopes two or more forms of an element that differ from each other according to their mass number

Lanthanides the elements in the periodic table with atomic numbers 58 through 71

Malleable capable of being hammered into thin sheets

Phosphor a material that gives off light when struck by electrons

Radioactive isotope an isotope that breaks apart and gives off some form of radiation

Rare earth elements *see* **Lanthanides**

Toxic poisonous

Occurrence in nature

Terbium is one of the rarest of the lanthanides. It ranks about 55th among the elements in the Earth's crust. It is about as abundant as **molybdenum** and **tungsten,** but more abundant than **iodine, silver,** and **gold.** Terbium occurs with other lanthanides in minerals such as monazite, cerite, gadolinite, xenotime, and euxenite.

Isotopes

Only one isotope of terbium occurs in nature, terbium-159. Isotopes are two or more forms of an element. Isotopes differ from each other according to their mass number. The number written to the right of the element's name is the mass number. The mass number represents the number of protons plus neutrons in the nucleus of an atom of the element. The number of protons determines the element, but the number of neutrons in the atom of any one element can vary. Each variation is an isotope.

Many radioactive isotopes of terbium are known also. A radioactive isotope is one that breaks apart and gives off some form of radiation. Radioactive isotopes are produced when very small particles are fired at atoms. These particles stick in the atoms and make them radioactive. The 17 radioactive isotopes of terbium carry mass numbers of 147 through 158 and 160 through 164.

One radioactive isotope, terbium-149, is used in medicine. The isotope is injected directly into cancer cells in a patient's body. The radiation given off kills the cancer cells. Terbium-149 is used because its radiation does not travel far, so it does not damage healthy cells. Therefore, it is safer to use than some other radioactive isotopes.

Extraction

The rare earth elements often occur together in the earth. A mineral like monazite may contain half a dozen or more different rare earth elements. The chemist's job, then, is to find a way to separate all these elements from each other.

Today, a standard procedure is available for separating the rare earth elements from each other. In this procedure, terbium usually ends up in the form of the compound terbium fluoride (TbF_3). The terbium may be obtained by passing an electric current through the compound:

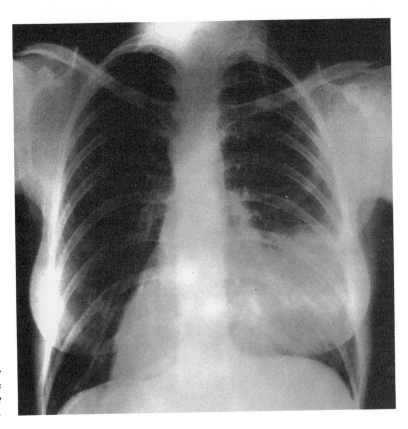

Terbium is often used in X-ray machines. This X ray shows pneumonia in the lower lobe of the patient's left lung.

$$2TbF_3 \rightarrow 2Tb + 3F_2$$

Reacting calcium metal with terbium fluoride also produces free terbium:

$$3Ca + 2TbF_3 \rightarrow 3CaF_2 + 2Tb$$

Uses and compounds

Probably the most common use of terbium and its compounds is in phosphors. A phosphor is a material that gives off light when struck by electrons. The back of a television screen is coated with different kinds of phosphors. When those phosphors are struck by electrons inside the television tube, they give off different colors of light. Phosphors that contain terbium give off a green light when struck by electrons. They are also used in X-ray screens to make very clear pictures.

Another use of terbium is in the manufacture of fuel cells. Any system that uses chemical reactions to produce electricity is a fuel cell. Fuel cells will probably be much more widely used as a source of electricity in the future. Terbium fuel cells operate effectively at very high temperatures.

Health effects

There is almost no information available on the health effects of terbium. In such cases, chemists use caution. The safest policy is to assume that terbium is very toxic and to avoid contact with it as much as possible.

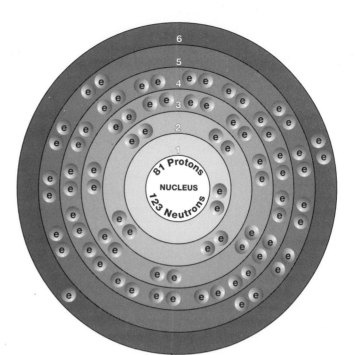

THALLIUM

Overview

Thallium is a member of the **aluminum** family, Group 13 (IIIA) on the periodic table. The periodic table is a chart that shows how chemical elements are related to one another. Thallium is also a member of the heavy metals, along with **gold, platinum,** and **lead.**

Thallium was first discovered by means of a spectroscope. A spectroscope is a device for analyzing the light produced when an element is heated. The spectrum (plural: *spectra*) of an element consists of a series of colored lines that are different for every element. The brightest lines in the spectrum of thallium are green, which accounts for its name. In Greek, the word *thallos* means "green twig." The green lines in thallium's spectrum look like green twigs.

Thallium is a rather uncommon element. Still, some of its compounds have important applications. For example, thallium sulfate (Tl_2SO_4) has long been used as a rodenticide (rat and mouse poison). One form of thallium is sometimes used to study the flow of blood in the body. It shows how well the heart is working.

SYMBOL
Tl

ATOMIC NUMBER
81

ATOMIC MASS
204.37

FAMILY
Group 13 (IIIA)
Aluminum

PRONUNCIATION
THA-lee-um

Discovery and naming

The spectroscope was invented in 1814 by German physicist Joseph von Fraunhofer (1787–1826). Forty years later, German chemists Robert Bunsen (1811–99) and Gustav Robert Kirchhoff (1824–87) improved on the instrument and showed how it could be used to study chemical elements. (See sidebar on Bunsen in the **cesium** entry in Volume 1.)

Scientists were fascinated by the instrument. They could detect the presence of elements without actually seeing them. A mineral is made of many elements, each of which gives off its own series of colored (spectral) lines. The spectroscope is able to detect all the elements present in the mineral.

Within a period of four years after the work of Bunsen and Kirchhoff, four new elements were discovered: **cesium, rubidium,** thallium, and **indium.** All four elements are named after the color of their spectral lines. The discoverer of thallium was British physicist Sir William Crookes (1832–1919).

Interestingly, thallium was discovered at almost the same time by French chemist Claude-Auguste Lamy (1820–78). Lamy discovered thallium the "old fashioned way," by separating one of its minerals in the laboratory. For a short time, there was a difference of opinion as to whether Lamy or Crookes was the "real" discoverer of thallium. Eventually, the decision was made in favor of Crookes.

Physical properties

Thallium is a heavy, bluish-white metal that resembles lead, element 82. Thallium is very soft and melts easily. It is soft enough to cut with an ordinary knife and will leave a mark on paper if rubbed across it.

Thallium has a melting point of 302°C (576°F) and a boiling point of 1,457°C (2,655°F). Its density is 11.85 grams per cubic centimeter.

Chemical properties

Thallium is a fairly active element. It reacts with acids and with **oxygen** in the air. When exposed to air, it forms a thin coating of thallium oxide (Tl_2O) that peels off easily. As the coating drops off, a new layer forms in its place.

WORDS TO KNOW

Alloy a mixture of two or more metals with properties different from those of the individual metals

Isotopes two or more forms of an element that differ from each other according to their mass number

Periodic table a chart that shows how chemical elements are related to each other

Radioactivity having the tendency to break apart and give off some form of radiation

Rodenticide a poison used to kill rats and mice

Spectroscope A device for analyzing the light produced when an element is heated

Superconductivity the tendency for an electric current to continue flowing through a material forever once it has begun

Toxic poisonous

Occurrence in nature

Thallium is quite uncommon in the Earth's crust. Its abundance is estimated to be about 0.7 parts per million. That puts it in the bottom half among the elements in terms of abundance. It is about as common as **iodine** or **tungsten.**

The most common minerals containing thallium are crookesite, lorandite, and hutchinsonite.

Isotopes

Two naturally occurring isotopes of thallium exist, thallium-203 and thallium-205. Isotopes are two or more forms of an element. Isotopes differ from each other according to their mass number. The number written to the right of the element's name is the mass number. The mass number represents the number of protons plus neutrons in the nucleus of an atom of the element. The number of protons determines the element, but the number of neutrons in the atom of any one element can vary. Each variation is an isotope.

Two dozen radioactive isotopes of thallium have also been made. A radioactive isotope is one that breaks apart and gives off some form of radiation. Radioactive isotopes are produced when very small particles are fired at atoms. These particles stick in the atoms and make them radioactive.

Thallium-201 is used by doctors to determine how well a person's heart is working. In many cases, the isotope is used as a stress test. Thallium-201 is injected into the patient's bloodstream as he or she exercises on a treadmill or bicycle. As soon as the exercise ends, the patient lies down. A large camera is passed over the body. The camera records the radiation given off by the isotope. This record shows whether the patient's heart is working properly or not.

Extraction

Thallium is obtained as a by-product of the recovery of lead and **zinc.** Gases from the recovery process are captured. They are then treated to obtain the pure metal.

Uses and compounds

For many years, thallium sulfate (Tl_2SO_4) was used as a rodenticide. It worked well with rats and mice because it passes through their skin easily. Once inside their bodies, it causes

 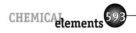

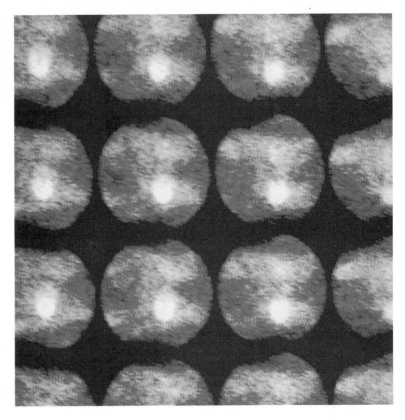

These images show the results of a thallium heart scan.

death. Thallium sulfate is also colorless and odorless, so rats and mice were not aware the compound was present.

Unfortunately, thallium sulfate has the same effects on humans. Accidental poisoning, especially of young children, led to the banning of thallium sulfate as a rodenticide in the United States in 1975. Today, safer compounds (for humans, not rats) are available for rodenticides.

Thallium is too expensive to have many practical applications. There are a few exceptions, however, that make use of special properties of the elements and its compounds. For example, thallium sulfide (Tl_2S) is sometimes used in photocells. Photocells are devices that convert light into electrical energy. In some kinds of light, thallium sulfide does not conduct electricity very well. But in other kinds of light, it conducts very well. Special photocells can be built to take advantage of this property.

An alloy of thallium and **mercury** can be used to make low-temperature thermometers. An alloy is made by melting and mixing two or more metals. The mixture has properties different from those of the individual metals. The thallium-mercury alloy remains liquid at –60°C (–76°F). At that temperature, a mercury-only thermometer would freeze solid.

An interesting new application of thallium is in superconducting materials. Superconductors have no resistance to the flow of electricity. Once an electrical current begins flowing in the material, it continues to flow forever. Superconducting materials may have some very important practical applications in the future.

Health effects
Both thallium and its compounds are very toxic. A person exposed to the element or its compounds over long periods of time develops weakness, pain in the arms and legs, and loss of hair. A high dosage in a short time leads to different symptoms. These symptoms include nausea, vomiting, diarrhea, pain in the arms and legs, coma, convulsions, and even death. People who work with thallium use extreme caution to avoid coming into contact with the material.

Accidental poisoning, especially of young children, led to the banning of thallium sulfate as a rodenticide in the United States in 1975.

THORIUM

Overview

Thorium is a member of the actinide family. The actinide elements are located in Row 7 of the periodic table. They have atomic numbers between 90 and 103. The periodic table is a chart that shows how chemical elements are related to one another. The actinide series is named for element 89, **actinium,** which is sometimes included in the actinide family.

Thorium was discovered in 1828 by Swedish chemist Jöns Jakob Berzelius (1779–1848). At the time, Berzelius did not realize that thorium was radioactive. That was discovered 70 years later, in 1898, by Polish-French physicist Marie Curie (1867–1934) and English chemist Gerhard C. Schmidt (1864–1949).

Thorium is a relatively common element with few commercial applications. There is some hope that it can someday be used in nuclear power plants, in which nuclear reactions are used to generate electricity.

Discovery and naming

In 1815, Berzelius was studying a new mineral found in the Falun district of Sweden. From his analysis, he concluded that

SYMBOL
Th

ATOMIC NUMBER
90

ATOMIC MASS
232.0381

FAMILY
Actinide

PRONUNCIATION
THOR-ee-um

he had found a new element. He named the element thorium, in honor of the Scandinavian god Thor.

Ten years later, Berzelius announced that he had made an error. The substance he had found was not a new element, but the compound **yttrium** phosphate (YPO_4).

Shortly thereafter, Berzelius again reported that he had found a new element. This time he was correct. He chose to retain thorium as the name for this element.

At the time Berzelius made his discovery, the concept of radioactivity was unknown. Radioactivity refers to the process by which an element spontaneously breaks down and gives off radiation. In that process, the element often changes into a new element. One of the first scientists to study radioactivity was Curie. She and Schmidt announced at almost the same time in 1898 that Berzelius' thorium was radioactive.

Physical properties

Thorium is a silvery white, soft, metal, somewhat similar to **lead.** It can be hammered, rolled, bent, cut, shaped, and welded rather easily. Its general physical properties are somewhat similar to those of lead. It has a melting point of about 1,800°C (3,300°F) and a boiling point of about 4,500°C (8,100°F). The density of thorium is about 11.7 grams per cubic centimeter.

Chemical properties

Thorium is soluble in acids and reacts slowly with **oxygen** at room temperature. At higher temperatures, it reacts with oxygen more rapidly, forming thorium dioxide (ThO_2).

Occurrence in nature

Thorium is a relatively abundant element in the Earth's crust. Scientists estimate that the crust contains about 15 parts per million of the element. That fact is important from a commercial standpoint. It means that thorium is much more abundant than another important radioactive element, **uranium.** Uranium is used in nuclear reactors to generate electricity and in making nuclear weapons (atomic bombs). Scientists believe thorium can replace uranium for these purposes. With more thorium than uranium available, it would be cheaper to make electricity with thorium than uranium.

Thorium in place of uranium?

Uranium is one of the most important elements in the world today. Why? One of its isotopes undergoes nuclear fission. Nuclear fission occurs when neutrons collide with the nucleus of a uranium atom. When that happens, the uranium nucleus splits apart. Enormous amounts of energy are released. That energy can be used for mass destruction in the form of atomic bombs, or used for peaceful energy production in nuclear power plants.

But there are two problems with using uranium for nuclear fission. First, of uranium's three isotopes (uranium-234, uranium-235, and uranium-238), only one—uranium-235—undergoes fission. The second problem is that this isotope of uranium is quite rare. Out of every 1,000 atoms of uranium, only seven are uranium-235. Tons of uranium ore must be processed and enriched to make tiny amounts of this critical isotope. It is difficult and extremely expensive.

Scientists know that another isotope of uranium, uranium-233, will also undergo fission. The problem is that uranium-233 does not occur in nature. So how can it be used to make atomic weapons or nuclear power?

The trick is to start with an isotope of thorium, thorium-232. Thorium-232 has a very long half life of 14 billion years. If thorium-232 is bombarded with neutrons, it goes through a series of nuclear changes, first to thorium-233, then to protactinium-233, and finally to uranium-233. The whole process only takes about a month. At the end of the month, a supply of uranium-233 has been produced. This isotope of uranium has a fairly long half life, about 163,000 years. So once it has been made, it stays around for a long time. It can then be used for nuclear fission.

Scientists would like to find a way to use this process to make uranium-233 economically. Thorium is much more abundant than uranium. It would be far cheaper to make nuclear bombs and nuclear power plants with thorium than with uranium.

Unfortunately, no one has figured how to make the process work on a large scale. One nuclear reactor using thorium was built near Platteville, Colorado, in 1979. However, a number of economic and technical problems developed. After only ten years of operation, the plant was shut down. The promise of thorium fission plants has yet to become reality.

The most common ores of thorium are thorite and monazite. Monazite is a relatively common form of beach sand. It can be found, among other places, on the beaches of Florida. This sand may contain up to 10 percent thorium.

Isotopes

More than two dozen isotopes of thorium are known. All are radioactive. The isotope with the longest half life is thorium-232. Its half life is about 14 billion years. Isotopes are two or more forms of an element. Isotopes differ from each other according to their mass number. The number written to the

There is some hope that thorium can someday be used in nuclear power plants, where nuclear reactions are used to generate electricity.

right of the element's name is the mass number. The mass number represents the number of protons plus neutrons in the nucleus of an atom of the element. The number of protons determines the element, but the number of neutrons in the atom of any one element can vary. Each variation is an isotope.

The half life of a radioactive element is the time it takes for half of a sample of the element to break down. After one half life (14 billion years), only 5 grams of a ten-gram sample of thorium-232 would be left. The remaining 5 grams would have broken down to form a new isotope.

Extraction

The thorium in monazite, thorite, or other minerals is first converted to thorium dioxide (ThO_2). This thorium dioxide is then heated with calcium to get the free element:

$$ThO_2 + 2Ca \rightarrow 2CaO + Th$$

Uses and compounds

Thorium and its compounds have relatively few uses. The most important thorium compound commercially is thorium dioxide. This compound has the highest melting point of any oxide, about 3,300°C (6,000°F). It is used in high-temperature ceramics. A ceramic is a material made from earthy materials, such as sand or clay. Bricks, tiles, cement, and porcelain are examples of ceramics. Thorium dioxide is also used in the manufacture of specialty glass and as a catalyst. A catalyst is a substance used to speed up or slow down a chemical reaction without undergoing any change itself.

The one device in which most people are likely to have seen thorium dioxide is in portable gas lanterns. These lanterns contain a gauzy material called a mantle. Gas passing through the mantle is ignited to produce a very hot, bright white flame. That flame provides the light in the lantern. The mantle in most lanterns was once made of thorium dioxide because it can get very hot without melting.

Opposite page:
Thorium fluoride produces the bright beam used in searchlights.

The thorium dioxide in a gas mantle is radioactive. But it is of no danger to people because the amount used is so small. Still, gas mantles in the United States are no longer made with thorium. Safer substitutes have been found.

The mantle in a portable gas lantern that produces a hot, white flame was once commonly made out of thorium dioxide. It was replaced, however, with safer substitutes, due to concerns of radioactivity.

Another thorium compound, thorium fluoride (ThF_4), is used in **carbon** arc lamps for movie projectors and searchlights. A carbon arc lamp contains a piece of carbon (charcoal) to which other substances (such as ThF_4) have been added. When an electric current is passed through the carbon, it gives off a bright white light. The presence of thorium fluoride makes this light even brighter.

Health effects

As with all radioactive materials, thorium is dangerous to the health of humans and other animals. It must be handled with great caution. Living cells that absorb radiation are damaged or killed. Inhaling a radioactive element is especially dangerous because it exposes fragile internal tissues.

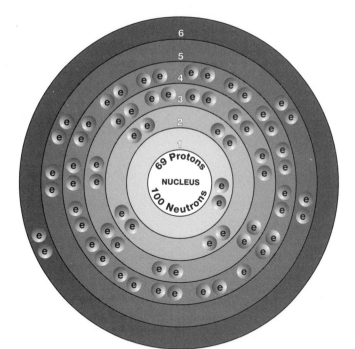

THULIUM

Overview

Thulium was given its name in honor of the earliest name for Scandanavia, Thule. The element was discovered and named by Swedish chemist Per Teodor Cleve (1840–1905) in 1879. Cleve made his discovery while studying the mineral erbia. Erbia was one of the many new elements found in a black rock discovered outside the town of Ytterby, Sweden, in 1787. The complete analysis of that rock took more than 100 years. In the process, nine new elements, including thulium, were discovered.

The chemical family to which thulium belongs is sometimes called the rare earth elements. That name is misleading. These elements are not really very rare. But they usually occur together in the earth, and they are very difficult to separate from each other. The more common chemical name for the group of rare earth elements is the lanthanides. The name comes from element 89, **lanthanum**, often classified as a lanthanide. The family makes up Row 6 of the periodic table. The periodic table is a chart that shows how chemical elements are related to each other.

Discovery and naming

In 1787, a Swedish army officer named Carl Axel Arrhenius (1757–1824) found a strange looking black rock outside the

SYMBOL
Tm

ATOMIC NUMBER
69

ATOMIC MASS
168.9342

FAMILY
Lanthanide
(rare earth metal)

PRONUNCIATION
THU-lee-um

town of Ytterby, Sweden. He passed it along to Johan Gadolin (1760–1852), professor of chemistry at the University of Åbo in Finland. Gadolin discovered a new mineral in the rock, now now known as yttria.

Gadolin had no idea how complicated yttria was. Other chemists worked for a hundred years before they completely understood the mineral.

In 1879, Cleve was studying one of the new elements found in yttria. The element had been named erbium by an earlier researcher. Cleve realized that erbium was not really an element, but was made up of three other substances. Cleve called these substances erbium, **holmium,** and thulium.

The substance Cleve called thulium was not pure thulium, but a compound of thulium combined with other elements. Pure thulium was not produced until 1910 by American chemist Charles James (1880–1928).

Physical properties

Thulium is a silvery metal so soft it can be cut with a knife. It is easy to work with and is both malleable and ductile. Malleable means capable of being hammered into thin sheets. Ductile means capable of being drawn into thin wires. Its melting point is 1,550°C (2,820°F) and its boiling point is 1,727°C (3,141°F). Its density is 9.318 grams per cubic centimeter.

Chemical properties

Thulium is relatively stable in air. That is, it does not react easily with **oxygen** or other substances in the air. It does react slowly with water and more rapidly with acids.

Occurrence in nature

Thulium compounds occur mixed with other rare earth compounds in minerals such as monazite, euxenite, and gadolinite. Monazite is about 0.007 percent thulium.

Thulium is probably the rarest of the lanthanide elements. Its abundance is estimated at about 0.2 to 1 part per million in the Earth's crust. This still makes it more abundant than **silver, platinum, mercury,** and **gold.**

WORDS TO KNOW

Ductile capable of being drawn into thin wires

Gamma rays a form of radiation similar to X rays

Isotopes two or more forms of an element that differ from each other according to their mass number

Lanthanides the elements in the periodic table with atomic numbers 58 through 71

Laser a device for making very intense light of one very specific color that is intensified many times over

Malleable capable of being hammered into thin sheets

Radioactive isotope an isotope that breaks apart and gives off some form of radiation

Rare earth elements *see* **Lanthanides**

Stable not likely to react with other materials

Toxic poisonous

Isotopes

Only one naturally occurring isotope of thulium exists, thulium-169. Isotopes are two or more forms of an element. Isotopes differ from each other according to their mass number. The number written to the right of the element's name is the mass number. The mass number represents the number of protons plus neutrons in the nucleus of an atom of the element. The number of protons determines the element, but the number of neutrons in the atom of any one element can vary. Each variation is an isotope.

At least 16 radioactive isotopes of thulium are known also. A radioactive isotope is one that breaks apart and gives off some form of radiation. Radioactive isotopes are produced when very small particles are fired at atoms. These particles stick in the atoms and make them radioactive.

Studies have been done on thulium-171, a radioactive isotope of thulium, for possible use in a portable X-ray machine. This

Thulium is a silvery metal so soft it can be cut with a knife.

isotope gives off gamma radiation. Gamma radiation is very similar to X rays. Gamma rays pass through soft tissues in the body like X rays. But they are blocked by bones and other thick materials. So a small amount of thulium-171 acts just like a tiny X-ray machine. It can be carried around more easily than can a big X-ray machine.

Extraction
Like many lanthanides, pure thulium is made by treating its fluorine compound with calcium:

$$2TmF_3 + 3Ca \rightarrow 3CaF_2 + 2Tm$$

Because it is so rare, thulium is also very expensive. It sells for about $50 a gram (about $23,000 a pound).

Uses
Thulium is too expensive to have many commercial uses. One of the few applications is in lasers. A laser is a device that produces bright, focused light of a single color. Thulium lasers work well at high temperatures and need less cooling. Lasers containing thulium are used in satellites that take pictures of the Earth's surface.

Compounds
There are no commercially important compounds of thulium.

Health effects
Although little is known about the health effects of thulium, it is assumed to be toxic. Caution is advised in handling or using it.

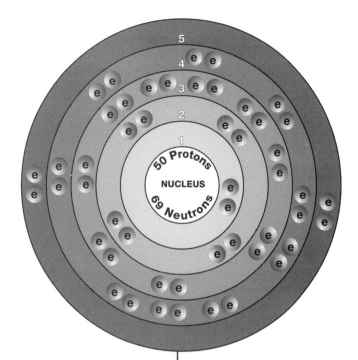

TIN

Overview

Tin is a member of Group 14 (IVA) in the periodic table. The periodic table is a chart that shows how chemical elements are related to one another. Tin is also part of the the carbon family. Other carbon family elements include **carbon, silicon, germanium,** and **lead.**

Tin is a highly workable metal that was once as valuable as **silver** for jewelry, coins, and special dishware. Today it is used as sheets in the construction of buildings and roofs, for soldering or joining metal parts, for storage containers, and in alloys like bronze and Babbitt metal.

Discovery and naming

Tin, its alloys, and its compounds have been known to humans for thousands of years. A number of references to the element can be found in the Bible. Tin was apparently known to other civilizations also. For example, the sacred Hindu book *Rig Veda,* written in about 1000 B.C., mentions tin among other metals known to the Hindus.

The alloy of tin known as bronze was probably produced even earlier than the pure metal. An alloy is made by melting and

SYMBOL
Sn

ATOMIC NUMBER
50

ATOMIC MASS
118.69

FAMILY
Group 14 (IVA)
Carbon

PRONUNCIATION
TIN

WORDS TO KNOW

Allotropes forms of an element with different physical and chemical properties

Bronze an alloy of copper and tin

Bronze Age a period in human history ranging from about 3500 B.C. to 1000 B.C., when bronze was widely used for weapons, utensils, and ornamental objects.

Cassiterite an ore of tin containing tin oxide, the major commercial source of tin metal

Ductile capable of being drawn into thin wires

Isotopes two or more forms of an element that differ from each other according to their mass number

Malleable capable of being hammered into thin sheets

Ore a compound or mixture from which an element can be extracted for commercial profit

Solder an alloy of tin and lead with a low melting point used to join two metals to each other

"Tin cry" a screeching-like sound made when tin metal is bent

Tinplate a type of metal consisting of thin protective coating of tin deposited on the outer surface of some other metal

mixing two or more metals. The mixture has properties that are different than any of the metals alone. The Egyptians, Mesopotamians, Babylonians, and Peruvians were producing bronze as far back as 2000 B.C. The alloy was probably discovered accidentally when copper and tin compounds were heated together. Over time, a method for producing consistent bronze was developed.

Bronze became popular among ancient peoples because it was harder and tougher than copper. Before the discovery of bronze, many metal items were made out of copper. But copper is soft and bends easily. Bronze is a much better replacement for copper in tools, eating utensils, and weapons. Bronze marked a significant advance in human civilization. This strong alloy improved transportation methods, food preparation, and quality of life during a period now known as the Bronze Age (4000–3000 B.C.).

The origin of the name tin is lost in history. Some scholars believe it is named for the Etruscan god Tinia. During the Middle Ages, the metal was known by its Latin name, *stannum*. It is from this name that the element's symbol, Sn, is derived.

Physical properties

The most common allotrope of tin is a silver-white metallic-looking solid known as the β-form (or "beta-form"). Allotropes are forms of an element with different physical and chemical properties. This "white tin" has a melting point of 232°C (450°F), a boiling point of 2,260°C (4,100°F), and a density of 7.31 grams per cubic centimeter.

One of tin's most interesting properties is its tendency to give off a strange screeching sound when it is bent. This sound is sometimes known as "tin cry." β-tin is both malleable and ductile. Malleable means capable of being hammered into thin sheets. Ductile means capable of being drawn into a thin wire. At temperatures greater than 200°C, tin becomes very brittle.

A second form of tin is α-tin (or "alpha-tin"), also known as "gray tin." Gray tin forms when white tin is cooled to temperatures less than about 13°C. Gray tin is a gray amorphous (lacking a crystalline shape) powder. The change from white tin to gray tin takes place rather slowly. This change is responsible for some peculiar and amazing changes in objects made from

 610 CHEMICAL **elements**

Tin sample.

the element. For example, tin and its alloys are used in jewelry, kitchenware, serving cups, and other metallic objects. When these objects are cooled below 13°C for long periods of time, the tin changes from a silvery, metallic material to a crumbly powder.

In the late nineteenth century, organ pipes in many cathedrals of Northern Europe were made of tin alloys. During the coldest winters, these pipes began to crumble as tin changed from one allotropic form to the other. The change was known as "tin disease." At the time, no one knew why this change occurred.

Chemical properties

Tin is relatively unaffected by both water and oxygen at room temperatures. It does not rust, corrode, or react in any other way. This explains one of its major uses: as a coating to protect other metals. At higher temperatures, however, the metal reacts with both water (as steam) and oxygen to form tin oxide.

One of tin's most interesting properties is its tendency to give off a strange screeching sound when it is bent. This sound is sometimes known as "tin cry."

Similarly, tin is attacked only slowly by dilute acids such as hydrochloric acid (HCl) and sulfuric acid (H_2SO_4). Dilute acids are mixtures that contain small amounts of acid dissolved in large amounts of water. This property also makes tin a good protective covering. It does not react with acids as rapidly as do many other kinds of metals, such as iron, and can be used, therefore, as a covering for those metals.

Tin dissolves easily in concentrated acids, however, and in hot alkaline solutions, such as hot, concentrated **potassium** hydroxide (KOH). The metal also reacts with the halogens to form compounds such as tin chloride and tin bromide. It also forms compounds with **sulfur, selenium,** and **tellurium.**

Occurrence in nature

Tin is not very abundant in nature. It ranks about 50th on the list of elements most commonly found in the Earth's crust. Estimates are that the crust contains about 1 to 2 parts per million of tin.

By far the most common ore of tin is cassiterite, a form of tin oxide (SnO_2). An ore is a compound or mixture from which an element can be extracted for commercial profit. Cassiterite has been mined for thousands of years as a source of tin. During ancient times, Europe obtained most of its tin from the British Isles. Today, the major producers of tin are China, Indonesia, Peru, Brazil, and Bolivia. The United States produces almost no tin of its own although it is the major consumer of the metal.

Isotopes

Tin has ten naturally occurring isotopes. Isotopes are two or more forms of an element. Isotopes differ from each other according to their mass number. The number written to the right of the element's name is the mass number. The mass number represents the number of protons plus neutrons in the nucleus of an atom of the element. The number of protons determines the element, but the number of neutrons in the atom of any one element can vary. Each variation is an isotope.

Fifteen radioactive isotopes have also been discovered. A radioactive isotope is one that breaks apart and gives off some form of radiation. Radioactive isotopes are produced when very small particles are fired at atoms. These particles stick in the atoms and make them radioactive.

None of the radioactive isotopes of tin have any commercial applications.

Extraction

Tin can be produced easily by heating cassiterite with charcoal (nearly pure **carbon**). In this reaction, the carbon reacts with and removes **oxygen** from the cassiterite, leaving pure tin behind.

$$SnO_2 + C \rightarrow Sn + CO_2$$

This reaction occurs so easily that people knew of the reaction thousands of years ago.

In order to obtain very pure tin, however, one problem must be solved. **Iron** often occurs in very small amounts along with tin oxide in cassiterite. Unless the iron is removed during the extraction process, a very hard, virtually unusable form of tin is produced. Modern systems of tin production, therefore, involve two steps. In one of those steps, impure tin is heated in the presence of oxygen to oxidize any iron in the mixture. In this reaction, iron is converted to iron(III) oxide, and metallic tin is left behind:

$$Sn + 4Fe + 3O_2 \rightarrow 2Fe_2O_3 + Sn$$

Uses

The largest amount of tin used in the United States goes to the production of solder. Solder is an alloy, usually made of tin and lead, with a low melting point. It is used to join two metals to each other. For example, metal wires are attached to electrical devices by means of solder. Solder is also used by plumbers to seal the joint between two metal pipes.

Solder is often applied by means of a soldering iron. A soldering iron consists of a steel bar through which an electric current runs. The electric current heats the bar as it passes through it. When a small piece of solder is placed on the tip of the soldering iron, it melts. The solder is then applied to the joint between two metals. When it cools, the bond is strong. In 1996, 15,600 metric tons of tin were used in the production of solder.

The largest amount of tin used in the United States goes to the production of solder.

Tin is also used in the manufacture of other alloys. Bronze, for example, is an alloy of tin and copper. In 1996, more than 2,750 metric tons of bronze were produced in the United States. It is used in a wide variety of industrial products, such as spark-resistant tools, springs, wire, electrical devices, water gauges, and valves.

One application of tin that was once important is in the manufacture of "tin foil." Tin foil is a very thin sheet of tin used to wrap candies, tobacco, and other products. The tin protected the products from spoiling by exposure to air. Today, most tin foil is actually thin sheets of **aluminum** because aluminum is less expensive.

Tin was once important in the manufacture of "tin foil." Now, aluminum is used because it is less expensive.

A very important application of tin is tinplating. Tinplating is the process by which a thin coat of tin is placed on the surface of steel, iron, or another metal. Tin is not affected by air, oxygen, water, acids, and bases to the extent that steel, iron, and other metals are. So the tin coating acts as a protective layer.

Perhaps the best known example of tin plating is in the production of food cans. Tin cans are made of steel and are covered with a thin layer of tin. Most food and drink cans today are made out of aluminum because it is cheaper.

Metals can be plated with tin in one of two ways. First, the metal to be plated can simply be dipped in molten (liquid) tin and then pulled out. A thin layer of liquid tin sticks to the base metal and then cools to form a thin coating. The second method is electroplating. In the process of electroplating, the base metal is suspended in a solution of tin sulfate, or a similar compound. An electric current passes through the solution, causing the tin in the solution to be deposited on the surface of the base metal.

Another tin alloy is Babbitt metal. Babbitt metal is a soft alloy made of any number of metals, including **arsenic,** cadmium, **lead,** or tin. Babbitt metal is used to make ball bearings for large industrial machinery. The Babbitt metal is laid down as a thin coating on heavier metal, such as iron or steel. The Babbitt metal retains a thin layer of lubricating oil more efficiently than iron or steel.

Tin toys

Most of today's toys are made of plastic. But until World War II (1939–45), most of the finest toys in the world were made of tin-plated metal. The earliest of these toys were made in the early 1800s. They were based on common objects and events, such as trains, horse-drawn carriages, sailing ships, and people from everyday life.

During the first half of the twentieth century, the most popular tin toy was the automobile. Toymakers made replicas of every type of car manufactured in the world. These toy cars ranged from the very simplest to the most detailed and elaborate.

Following World War II, plastic toys became much more popular, but tin toys were still made. Reflecting the times, these toys often represented space ships, robots, and other modern objects.

The manufacture of tin toys is no longer the large-scale business it was one hundred years ago. However, toy-collecting has remained a fascinating hobby for adults and children around the world. Some antique collectors and dealers specialize in tin toys.

Compounds

About a sixth of all tin consumed in the United States is used in the production of tin compounds. Some of the most important of those compounds and their uses are as follows:

tin chloride ($SnCl_2$): used in the manufacture of dyes, polymers, and textiles; in the silvering of mirrors; as a food preservative; as an additive in perfumes used in soaps; and as an anti-gumming agent in lubricating oils

tin oxide (SnO_2): used in the manufacture of special kinds of glass, ceramic glazes and colors, perfumes and cosmetics, and textiles; and as a polishing material for steel, glass, and other materials

tin chromate ($SnCrO_4$ or $Sn(CrO_4)_2$): brown or yellowish-brown compounds used as a coloring agent for porcelain and china

tin fluoride (SnF_2) and tin pyrophosphate ($Sn_2P_2O_7$): used as toothpaste additives to help protect against cavities

Health effects

Most compounds of tin are toxic (poisonous). Tin compounds are most likely to present a hazard when they get into the air. Then, they may be inhaled, after which they can cause problems such as nausea, diarrhea, vomiting, and cramps.

The U.S. government has set a standard of 2 milligrams per cubic meter of air for most tin compounds. For organic compounds of tin (those that contain the element carbon also), the limit is 0.1 milligram per cubic meter. Miners and factory workers are the most likely people to be exposed to these levels of tin. The amount of tin absorbed from canned foods is too small to be of concern to consumers.

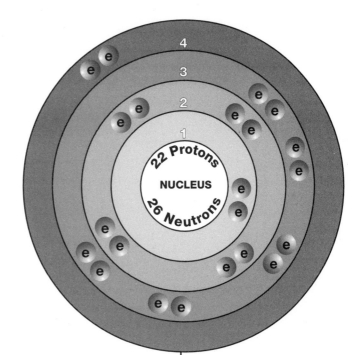

TITANIUM

Overview

Titanium is found in the middle of the periodic table. The periodic table is a chart that shows how chemical elements are related to one another. Titanium is a transition metal and is part of Group 4 (IVB).

Titanium was one of the first elements to be discovered by modern chemists. The "modern" chemistry period begins after the middle of the eighteenth century. That period is chosen because it is the first time that the basic concepts of modern chemistry were developed.

Titanium was discovered by English clergyman William Gregor (1761–1817). Gregor studied minerals as a hobby. He did not think of himself as a chemist, and yet his research led to the discovery of titanium.

Titanium and its compounds have become very important in modern society. The metal is widely used in a variety of alloys. An alloy is made by melting and mixing two or more metals. The mixture has properties different from those of the individual metals. Titanium alloys are used in aircraft, spacecraft, jewelry, clocks, armored vehicles, and in the construction of buildings.

SYMBOL
Ti

ATOMIC NUMBER
22

ATOMIC MASS
47.88

FAMILY
Group 4 (IVB)
Transition metal

PRONUNCIATION
ty-TAY-nee-um

Alloy a mixture of two or more metals with properties different from those of the individual metals

Biocompatible not causing a reaction when placed into the body

Catalyst a substance used to speed up or slow down a chemical reaction without undergoing any change itself

Corrosive agent a material that tends to vigorously react or eat away at something

Ductile capable of being drawn into thin wires

Isotopes two or more forms of an element that differ from each other according to their mass number

Malleable capable of being hammered into thin sheets

Periodic table a chart that shows how chemical elements are related to one another

Radioactive isotope an isotope that breaks apart and gives off some form of radiation

Slag a mixture of materials that separates from a metal during its purification and floats on top of the molten metal

Discovery and naming

Gregor discovered titanium while he was studying a mineral found near his home. He was able to identify most of the mineral, but he found one part that he could not identify. He decided it was a new substance, but did not continue his research. Instead, he wrote a report and left it to professional chemists to find out more about the material.

Today, we know that the material Gregor found is a mineral called ilmenite. Ilmenite is made of **iron, oxygen,** and titanium. Its chemical formula is $FeTiO_3$. Even though Gregor did not complete his study of ilmenite, he is usually given credit for the discovery of titanium.

Surprisingly, most chemists paid little attention to Gregor's report. Four years later, German chemist Martin Heinrich Klaproth (1743–1817) decided to study ilmenite. Klaproth believed that Gregor had been correct and that ilmenite truly did contain a new element. Klaproth suggested the name titanium, in honor of the Titans. The Titans were mythical giants who ruled the Earth until they were overthrown by the Greek gods. Klaproth reminded everyone that Gregor should receive credit for having discovered the element.

Klaproth was never able to produce pure titanium from ilmenite, only titanium dioxide (TiO_2). It was not until 1825 that even impure titanium metal was produced. Swedish chemist Jöns Jakob Berzelius (1779–1848) accomplished this task.

Physical properties

Pure titanium metal can exist as a dark gray, shiny metal or as a dark gray powder. It has a melting point of 1,677°C (3,051°F) and a boiling point of 3,277°C (5,931°F). Its density is 4.6 grams per cubic centimeter. Titanium metal is brittle when cold and can break apart easily at room temperature. At higher temperatures, it becomes malleable and ductile. Malleable means capable of being hammered into thin sheets. Ductile means capable of being drawn into thin wires.

Titanium has an interesting physical property. Small amounts of oxygen or **nitrogen,** make it much stronger.

Chemical properties

In general, titanium tends to be quite unreactive. It does not combine with oxygen at room temperature. It also resists attack by acids, **chlorine,** and other corrosive agents. A corrosive agent is a material that tends to vigorously react or eat away at something.

Titanium becomes more reactive at high temperatures. It can actually catch fire when heated in the presence of oxygen.

Occurrence in nature

Titanium is a very common element. It is the ninth most abundant element in the Earth's crust. Its abundance is estimated to be about 0.63 percent. That places titanium just above hydrogen and just below **potassium** among elements present in the earth.

The most common mineral sources of titanium are ilmenite, rutile, and titanite. Titanium is also obtained from **iron** ore slags. Slag is an earthy material that floats to the top when iron is removed from iron ore.

Titanium alloys are used in clocks.

Titanium metal is brittle when cold and can break apart easily at room temperature.

Titanium products at a mill.

Isotopes

Five naturally occurring isotopes of titanium exist. They are titanium-46, titanium-47, titanium-48, titanium-49, and titanium-50. The most abundant of these is titanium-48. It makes up about 75 percent of all titanium found in nature. Isotopes are two or more forms of an element. Isotopes differ from each other according to their mass number. The number written to the right of the element's name is the mass number. The mass number represents the number of protons plus neutrons in the nucleus of an atom of the element. The number of protons determines the element, but the number of neutrons in the atom of any one element can vary. Each variation is an isotope.

Four artificial isotopes of titanium have also been made. These are all radioactive. A radioactive isotope is one that breaks apart and gives off some form of radiation. Radioactive isotopes are produced when very small particles are fired at atoms. These particles stick in the atoms and make them radioactive.

None of the radioactive isotopes of titanium has any commercial applications.

Extraction

The methods used to obtain titanium are similar to those used for other metals. One way to make the metal is to heat one of its compounds with another metal, such as **magnesium:**

$$2Mg + TiCl_4 \rightarrow Ti + 2MgCl_2$$

Another approach is to pass an electric current through a molten (melted) compound of titanium:

$$TiCl_4 \xrightarrow{\text{electricity}} Ti + 2Cl_2$$

Uses

By far the most important use of titanium is in making alloys. The metal is most commonly added to steel. It adds strength to the steel and makes it more resistant to corrosion (rusting). Titanium also has another advantage in alloys. Its density is less than half that of steel. So a steel alloy containing titanium weighs less, pound-for-pound, than does the pure steel alloy.

These properties explain why titanium steel is so desirable for spacecraft and aircraft applications. In fact, about 65 percent of all titanium sold is used in aerospace applications. Titanium alloys are used in the airframes (bodies) and engines of aircraft and spacecraft. Other uses are in armored vehicles, armored vests, and helmets, in jewelry, eyeglasses, bicycles, golf clubs, and other sports equipment; in specialized dental implants; in power-generating plants and other types of factories; and in roofs, faces, columns, walls, ceilings, and other parts of buildings.

Titanium alloys have also become popular in body implants, such as artificial hips and knees. These alloys are light, strong, long-lasting, and biocompatible. Biocompatible means that the alloy does not cause a reaction when placed into the body.

Compounds

The most important compound of titanium is titanium dioxide (TiO_2). In 1996, 1,230,000 metric tons of this compound was produced in the United States. Titanium dioxide is a dense white powder with excellent hiding power. That term means that anything beneath it cannot be seen well. This property accounts for the major use of titanium dioxide: making white

Titanium tetrachloride combines with moisture in the air to form a dense white cloud. Skywriters use titanium tetrachloride to form letters in the sky.

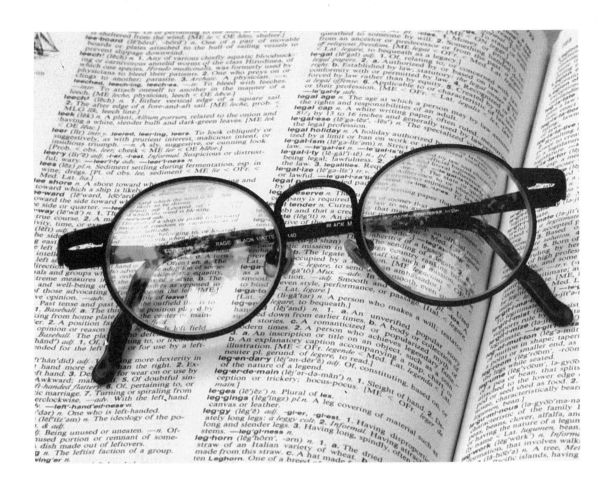

paint. Titanium dioxide paint is a good choice for painting over old wallpaper or dark paints because it covers so well. In 1996, about half of the titanium dioxide produced in the United States was used in paints.

About 40 percent of all titanium dioxide used in the United States goes into paper and plastic materials. Titanium dioxide gives "body" to paper and makes it opaque (unable to see through it). Other uses are in floor coverings, fabrics and textiles, ceramics, ink, roofing materials, and catalysts in industrial operations. A catalyst is a substance used to speed up or slow down a chemical reaction without undergoing any change itself.

Another interesting compound is titanium tetrachloride ($TiCl_4$). Titanium tetrachloride is a clear, colorless liquid when kept in a sealed container. However, it changes dramatically when exposed to the air. It combines with moisture in the air to form a dense white cloud. Skywriters use titanium tetrachloride to form letters in the sky. The compound is also used to make smokescreens. Smoke effects used in motion pictures and television programs sometimes are produced with titanium tetrachloride.

Health effects

Titanium appears to have no harmful effects on plants or humans. It has also not been shown to have any role in maintaining good health.

TRANSFERMIUM ELEMENTS

Overview

The term "transfermium" describes the elements with atomic numbers greater than 100. **Fermium** is element 100, so *transfermium* means "beyond fermium." The transfermium elements are grouped together for a number of reasons. First, they are all prepared artificially. None of them occur in the Earth's crust naturally (that anyone knows of). Second, they can be made with only the greatest difficulty. In fact, no more than a few atoms of some transfermium elements have been created so far. Third, very little is known about the transfermium elements. With only a few atoms to study, it is difficult to learn much about them.

Still, the transfermium elements are of great interest to chemists and physicists. They help answer questions about the periodic table. The periodic table is a chart that shows how chemical elements are related to each other. The transfermium elements are found at the very end of the periodic table. Scientists want to know if there is a limit to how heavy a chemical element can be. They also want to know what the properties of these very heavy elements will be like.

The chart below gives basic information about the transfermium elements. A discussion of the names and symbols in the chart follows in the next section.

Name	Atomic Symbol	Atomic Number	Mass
Mendelevium	Md	101	258
Nobelium	No	102	259
Lawrencium	Lr	103	260
Rutherfordium	Rf	104	261
Dubnium	Db	105	262
Seaborgium	Sg	106	263
Bohrium	Bh	107	262
Hassium	Hs	108	265
Meitnerium	Mt	109	266
Ununnilium	Uun	110	–
Unununium	Uuu	111	–
Ununbiium	Uub	112	–

WORDS TO KNOW

Half life the time it takes for half of a sample of a radioactive element to break down

Isotopes two or more forms of an element that differ from each other according to their mass number

"Magic number" the number of protons and/or neutrons in an atom that tend to make the atom stable (not radioactive)

Particle accelerator ("atom smasher") a machine that makes very tiny particles, like protons or small atoms, move very fast

Periodic table a chart that shows how chemical elements are related to each other

Transfermium element any element with an atomic number greater than 100

Discovery of the elements

All transfermium elements are made in particle accelerators, or "atom smashers." A particle accelerator is a machine that makes very tiny particles, like protons or small atoms, move very fast. They often go nearly as fast as the speed of light. Light travels about 300,000,000 meters per second (186,000 miles per second).

These fast moving particles are then made to smash into atoms. If they hit an atom just right, they will stick to the atom, making it heavier. For example, when fast moving **neon** atoms strike atoms of **americium,** the following reaction can occur:

$$\,_{95}^{243}\text{Am} + \,_{10}^{22}\text{Ne} \rightarrow \,_{105}^{265}\text{Db}$$

The new element, dubnium (number 105), is produced.

This kind of experiment is easy to describe but very difficult to carry out. In fact, this research is carried out at only three laboratories in the world. One is the Joint Institute of Nuclear Research, in Dubna, Russia. The second is the Lawrence Berkeley Laboratory at the University of California at Berkeley in the United States. The third is the Institute for Heavy Ion

Two men in a particle accelerator.
This one is in Strasbourg, France.

Research in Darmstadt, Germany. All three laboratories use large particle accelerators that cost millions of dollars. Dozens of scientists from many different countries work on each team.

Credit for discovery of a transfermium element is extremely complicated. In most cases, no more than a handful of atoms is produced in an atom smasher. For example, the Dubna group first claimed to have found element 104 in 1964, but many scientists doubted this report. Five years later, American scientists also reported making element 104. This time, the evidence was better.

Naming the elements

One reason that scientists often argue over the discovery of an element is this: The group of scientists that discovers an element usually has the opportunity to suggest a name for it. For example, researchers at the Berkeley laboratory first discovered elements 97 and 98. They suggested naming those elements

berkelium and **californium,** in honor of Berkeley, California, where the research was done.

The final decision about naming elements is made by a group called the International Union of Pure and Applied Chemistry (IUPAC). The decision can take a very long time. The IUPAC spent nearly 20 years trying to agree on names for elements 104, 105, and 106. Finally, in 1997, the IUPAC announced the official and final names for elements 101 through 109. Those names and their symbols are shown in the accompanying chart.

The names chosen by the IUPAC honor either great scientists or places of importance. The meaning of the names is as follows:

> Mendelevium (Md): named after Russian chemist Dmitri Mendeleev (1834–1907), who developed the periodic law and the periodic table
>
> Nobelium (No): named after Swedish inventor Alfred Nobel (1833–96), who provided funding for the Nobel Prizes when he died
>
> Lawrencium (Lr): named after American physicist Ernest Orlando Lawrence (1901–58), who invented one of the first particle accelerators and for whom the Lawrence Berkeley Laboratory is named
>
> Rutherfordium (Rf): named after British physicist Ernest Rutherford (1871–1937), who made many important discoveries about atoms and radioactivity
>
> Dubnium (Db): named after Dubna, the city in Russia where the Joint Institute for Nuclear Research is located
>
> Seaborgium (Sg): named after American chemist Glenn Seaborg (1912–), who has been involved in the discovery of ten elements
>
> Bohrium (Bh): named after Danish physicist Niels Bohr (1885–1962), who helped develop the modern theory of the atom

The group of scientists that discovers an element usually has the opportunity to suggest a name for it.

*Ernest O. Lawrence, in front of a
particle-accelerating cyclotron.
Lawrencium is named after him.*

Hassium (Hs): named after the German state in
which the Institute for Heavy Ion Research is located

Meitnerium (Mt): named after Austrian physicist Lise
Meitner (1878–1968), who helped explain the
process of nuclear fission (the splitting of atoms)

The IUPAC has not yet assigned names for elements 110
through 112. Discovery of those elements has been announced
by the German team. But those discoveries have not yet been

confirmed. Therefore, temporary names have been assigned. Those names come from the Latin words for the numbers 110, 111, and 112. They are ununnilium (Uun), unununium (Uuu), and ununbiium (Uub).

Properties of the elements

No one knows much about the properties of the transfermium elements. It isn't possible to see or touch or smell or taste any of these elements. There are often no more than a few dozen atoms to study.

In fact, it is quite amazing that scientists know much of anything about these elements. Yet, they do know a few things. In 1997, for example, the German team studied the properties of element 106, seaborgium, with only six atoms to work with! But they managed to watch how these atoms behaved as they slowly moved down a column of special material.

Of course, these elements have no uses.

Isotopes

Most transfermium elements have more than one isotope. Isotopes are two or more forms of an element. Isotopes differ from each other according to their mass number. The number written to the right of the element's name is the mass number. The mass number represents the number of protons plus neutrons in the nucleus of an atom of the element. The number of protons determines the element, but the number of neutrons in the atom of any one element can vary. Each variation is an isotope.

The number of isotopes currently known for each element follows. The numbers may change as scientists discover new isotopes.

mendelevium: 13
nobelium: 11
lawrencium: 8
rutherfordium: 6
dubnium: 5
seaborgium: 3
bohrium: 2
hassium: 1
meitnerium: 1

An island of stability?

The larger an atom is, the more unstable it tends to be. It seems that big atoms have trouble staying together. They tend to fall apart, giving off tiny particles like electrons and protons in the process. When they do so, they change into other, smaller atoms. This process is called radioactive decay.

All elements heavier than bismuth are radioactive. They have no stable isotopes. Does that mean that scientists will never find another stable element in the transfermium group? As they search for elements 110, 111, 112, and beyond, will they always find radioactive isotopes only?

Some scientists think the answer is no. They believe that some very heavy elements may be stable. Their atoms may be able to stay together, as is the case with lighter elements. One of these elements may be number 114.

Scientists think that atoms are likely to be stable if they contain a certain "magic" number of protons and neutrons. Those magic numbers are 2, 8, 20, 28, 50, 82, 114, and 184. So an atom with 20 protons and 20 neutrons, for example, would be expected to be stable. And it is.

The next element among the transfermium elements with a "magic" number of protons and neutrons is number 114. One isotope of element 114 could have 114 protons and 184 neutrons. It would have a "double magic" number and might be very stable. Scientists hope to be able to make enough of the element to study. The element might or might not have practical uses. But it would be an exciting discovery for scientists.

All of the isotopes of the transfermium elements are radioactive. A radioactive isotope is one that breaks apart and gives off some form of radiation. In most cases, they have very short half lives. The half life of a radioactive isotope is the time it takes for half of a sample to break apart.

The half life of most transfermium isotopes is only a few seconds or less. The half life of dubnium-260, for example, is 1.6 seconds. That means that half of the atoms in a sample will break down in 1.6 seconds and change to some other element. Short half lives of the transfermium isotopes makes them hard to study. They tend to break down almost as soon as they are formed. Scientists have very little time to observe them.

TUNGSTEN

Overview

Tungsten is a transition metal. The transition metals are a group of elements found in the middle of the periodic table. They occupy the boxes in Rows 4 through 7 between Groups 2 and 13. The periodic table is a chart that shows how chemical elements are related to one another.

These metals have very similar physical and chemical properties. One of tungsten's unusual properties is its very high melting point of 3,410°C (6,170°F). This is the highest melting point of any metal. Another of its important properties is its ability to retain its strength at very high temperatures. These properties account for tungsten's primary application, the manufacture of alloys. An alloy is made by melting and mixing two or more metals. The mixture has properties different from those of the individual metals.

Credit for the discovery of tungsten is often divided among three men—Spanish scientists Don Fausto D'Elhuyard (1755–1833) and his brother Don Juan José D'Elhuyard (1754–96), and Swedish chemist Carl Wilhelm Scheele (1742–86). Tungsten's chemical symbol, W, is taken from an alternative name for the element, wolfram.

SYMBOL
W

ATOMIC NUMBER
74

ATOMIC MASS
183.85

FAMILY
Group 6 (VIB)
Transition metal

PRONUNCIATION
TUNG-stun

Discovery and naming

The first mention of tungsten and its compounds can be traced to about 1761. German chemist Johann Gottlob Lehmann (1719–67) was studying a mineral known as wolframite. He found two new substances in the mineral but did not recognize that they were new elements.

About twenty years later, Scheele also studied this mineral. He produced from it a white acidic powder. Scheele knew the powder was a new substance. But he could not isolate a pure element from it. Scheele's discovery was actually tungstic acid (H_2WO_4). (See sidebar on Scheele in the **chlorine** entry in Volume 1.)

Tungsten metal was prepared for the first time in 1783 by the D'Elhuyard brothers. In 1777, they were sent to Sweden to study mineralogy. After their return to Spain, the brothers worked together on a number of projects. One project involved an analysis of wolframite. They produced tungstic acid like Scheele but went one step further. They found a way to obtain pure tungsten metal from the acid. For this work, they are generally given credit as the discoverers of tungsten.

The name tungsten is taken from the Swedish phrase that means "heavy stone." In some parts of the world, the element is still called by another name, wolfram. This name comes from the German expression Wolf rahm, or "wolf froth (foam)." The element's chemical symbol is taken from the German name rather than the Swedish name.

Physical properties

Tungsten is a hard brittle solid whose color ranges from steel-gray to nearly white. Its melting point is the highest of any metal, 3,410°C (6,170°F) and its boiling point is about 5,900°C (10,600°F). Its density is about 19.3 grams per cubic centimeter. Tungsten conducts electrical current very well.

Chemical properties

Tungsten is a relatively inactive metal. It does not combine with **oxygen** at room temperatures. It does corrode (rust) at temperatures above 400°C (700°F). It does not react very readily with acids, although it does dissolve in nitric acid or aqua regia. Aqua regia is a mixture of hydrochloric and nitric

WORDS TO KNOW

Alloy a mixture of two or more metals with properties different from those of the individual metals

Isotopes two or more forms of an element that differ from each other according to their mass number

Periodic table a chart that shows how chemical elements are related to each other

Radioactive isotope an isotope that breaks apart and gives off some form of radiation

Transition metal an element in Groups 3 through 12 of the periodic table

acids. It often reacts with materials that do not react with either acid separately.

Tungsten samples.

Occurrence in nature

Tungsten never occurs as a free element in nature. Its most common ores are the minerals scheelite, or calcium tungstate ($CaWO_4$) and wolframite, or **iron manganese** tungstate ($(Fe,Mn)WO_4$). The abundance of tungsten in the Earth's crust is thought to be about 1.5 parts per million. It is one of the more rare elements.

The largest producers of tungsten in the world are China, Russia, and Portugal. No tungsten was mined in the United States in 1996. Detailed information about the production and use of tungsten in the United States is not available. This information is withheld from the public to protect the companies that produce and use tungsten.

Isotopes

Five naturally occurring isotopes of tungsten exist. They are tungsten-180, tungsten-182, tungsten-183, tungsten-184, and tungsten-186. Isotopes are two or more forms of an element. Isotopes differ from each other according to their mass

In some parts of the world, tungsten is still called by another name, wolfram. This name comes from the German expression *Wolf rahm*, or "wolf froth (foam)."

number. The number written to the right of the element's name is the mass number. The mass number represents the number of protons plus neutrons in the nucleus of an atom of the element. The number of protons determines the element, but the number of neutrons in the atom of any one element can vary. Each variation is an isotope.

About a dozen radioactive isotopes of tungsten are known also. A radioactive isotope is one that breaks apart and gives off some form of radiation. Radioactive isotopes are produced when very small particles are fired at atoms. These particles stick in the atoms and make them radioactive.

None of the radioactive isotopes of tungsten has any important commercial use.

Extraction

Tungsten metal can be obtained by heating tungsten oxide (WO_3) with **aluminum:**

$$2Al + WO_3 \rightarrow W + Al_2O_3$$

It also results from passing **hydrogen** gas over hot tungstic acid (H_2WO_4):

$$H_2WO_4 + 3H_2 \rightarrow W + 4H_2O$$

Uses

By far the most important use of tungsten is in making alloys. Tungsten is used to increase the hardness, strength, elasticity (flexibility), and tensile strength (ability to stretch) of steels. The metal is usually prepared in one of two forms. Ferrotungsten is an alloy of iron and tungsten. It usually contains about 70 to 80 percent tungsten. Ferrotungsten is mixed with other metals and alloys (usually steel) to make specialized alloys. Tungsten is also produced in powdered form. It can then be added to other metals to make alloys.

About 90 percent of all tungsten alloys are used in mining, construction, and electrical and metal-working machinery. These alloys are used to make high-speed tools; heating elements in furnaces; parts for aircraft and spacecraft; equipment used in radio, television, and radar; rock drills; metal-cutting tools; and similar equipment.

About 90 percent of all tungsten alloys are used in mining, construction, and electrical and metal-working machinery.

A small, but very important, amount of tungsten is used to make incandescent lights. The very thin metal wire that makes up the filament in these lights is made of tungsten. An electric current passes through the wire, causing it to get hot and give off light. It does not melt because of the high melting point of tungsten.

Compounds

Probably the most important compound of tungsten is tungsten carbide (WC). Tungsten carbide has a very high melting point of 2,780°C (5,000°F). It is the strongest structural material. It is used to make parts for electrical circuits, cutting tools, cermets, and cemented carbide. A cermet is a material made of a ceramic and a metal. A ceramic is a clay-like material. Cermets are used where very high temperatures occur for long periods of time. For example, the parts of a rocket motor or a jet engine may be made from a cermet.

Tungsten alloys are used in radar equipment. Here, Doppler radar measures the speed and direction of local winds.

A cemented carbide is made by bonding tungsten carbide to another metal. The product is very strong and remains strong at high temperatures. Cemented carbides are used for rock and metal cutting. They can operate at 100 times the speed of similar tools made of steel.

Health effects

Tungsten has no essential role in the health of plants, humans, or animals. In moderate amounts, it also presents virtually no health danger.

URANIUM

Overview

Uranium is the heaviest and last naturally occurring element in the periodic table. The periodic table is a chart that shows how chemical elements are related to each other. Uranium occurs near the beginning of the actinide family. The actinide family consists of elements with atomic numbers 90 through 103.

At one time, uranium was considered to be a relatively unimportant element. It had a few applications in the making of stains and dyes, in producing specialized steels, and in lamps. But annual sales before World War II (1939–45) amounted to no more than a few hundred metric tons of the metal and its compounds.

Then, a dramatic revolution occurred. Scientists discovered that one form of uranium will undergo nuclear fission. Nuclear fission is the process in which the nuclei of large atoms break apart. Large amounts of energy and smaller atoms are produced during fission. The first application of this discovery was in the making of nuclear weapons, such as the atomic bomb. After the war, nuclear power plants were built to make productive use of nuclear fission. Nuclear power plants convert the energy released by fission to electricity. Today, uranium is

SYMBOL
U

ATOMIC NUMBER
92

ATOMIC MASS
238.0289

FAMILY
Actinide

PRONUNCIATION
yUH-RAY-nee-um

WORDS TO KNOW

Actinide family elements with atomic numbers 90 through 103

Ductile capable of being drawn into thin wires

Half life the time it takes for half of a sample of a radioactive element to break down

Isotopes two or more forms of an element that differ from each other according to their mass number

Malleable capable of being hammered into thin sheets

Mordant a material that helps a dye stick to cloth

Nuclear fission a process in which neutrons collide with the nucleus of a uranium atom causing it to split apart with the release of very large amounts of energy

Periodic table a chart that shows how chemical elements are related to each other

Radioactive isotope an isotope that breaks apart and gives off some form of radiation

regarded as one of the most important elements for the future of the human race.

Discovery and naming

Credit for the discovery of uranium is usually given to German chemist Martin Klaproth (1743–1817). During the late 1780s, Klaproth was studying a common and well-known ore called pitchblende. At the time, scientists thought that pitchblende was an ore of **iron** and **zinc.**

During his research, however, Klaproth found that a small portion of the ore did not behave the way iron or zinc would be expected to behave. He concluded that he had found a new element and suggested the name uranium for the element. The name was given in honor of the Uranus, a planet that had been discovered only a few years earlier, in 1781.

For some time, scientists believed that Klaproth had isolated uranium. Eventually they realized he had found uranium oxide (UO_2), a compound of uranium. It was not until a half century later, in fact, that the pure element was prepared. In 1841, French chemist Eugène-Melchior Peligot (1811–90) produced pure uranium from uranium oxide.

Early researchers did not know that uranium was radioactive. In fact, radioactivity was not discovered until 1898. Radioactivity is the tendency of an isotope or element to break down and give off radiation.

Physical properties

Uranium is a silvery, shiny metal that is both ductile and malleable. Ductile means capable of being drawn into thin wires. Malleable means capable of being hammered into thin sheets. Its melting point is 1,132.3°C (2,070.1°F) and its boiling point is about 3,818°C (6,904°F). Its density is about 19.05 grams per cubic centimeter.

Chemical properties

Uranium is a relatively reactive element. It combines with nonmetals such as **oxygen, sulfur, chlorine, fluorine, phosphorus,** and **bromine.** It also dissolves in acids and reacts with water. It forms many compounds that tend to have yellowish or greenish colors.

Occurrence in nature

Uranium is a moderately rare element. Its abundance is estimated to be about 1 to 2 parts per million, making it about as
abundant as bromine or **tin.** The most common ore of uranium
is pitchblende, although it also occurs in other minerals, such
as uraninite, carnotite, uranophane, and coffinite.

Isotopes

All isotopes of uranium are radioactive. Three of these occur
naturally, uranium-234, uranium-235, uranium-238. By far the
most common is uranium-238, making up about 99.276% of
uranium found in the Earth's crust. Uranium-238 also has the
longest half life, about 4,468,000,000 years.

Isotopes are two or more forms of an element. Isotopes differ
from each other according to their mass number. The number
written to the right of the element's name is the mass number.
The mass number represents the number of protons plus neutrons

*A pitchblende ore (uranium)
sample.*

in the nucleus of an atom of the element. The number of protons determines the element, but the number of neutrons in the atom of any one element can vary. Each variation is an isotope.

The half life of a radioactive element is the time it takes for half of a sample of the element to break down. Imagine that the Earth's crust contains 100 million tonnes of uranium-238 today. Only half of a uranium-238 sample would remain 4,468,000,000 years from now (one half life). The remainder would have changed into other isotopes.

About a dozen other isotopes of uranium have been made artificially.

Extraction

Uranium is mined in much the same way iron is. Ore is removed from the earth, then treated with nitric acid to make uranyl nitrate ($UO_2(NO_3)_2$). This compound is converted to ura-

A nuclear explosion at sea. The explosion occurs as a result of uranium-235 undergoing nuclear fission.

nium dioxide (UO_2). Finally, this compound is converted to pure uranium metal with **hydrogen** gas:

$$UO_2 + 2H_2 \rightarrow 2H_2O + U$$

Uses and compounds

Uranium compounds have been used to color glass and ceramics for centuries. Scientists have found that glass made in Italy as early as A.D. 79 was colored with uranium oxide. They have been able to prove that the coloring was done intentionally.

Some uranium compounds were used for this purpose until quite recently. In fact, a popular type of dishware known as "Fiesta Ware" made in the 1930s and 1940s sometimes used uranium oxide as a coloring material. Other glassware, ceramics, and glazes also contained uranium oxide as a coloring agent.

Uranium compounds have had other limited uses. For example, they have been used as mordants in dyeing operations. A mor-

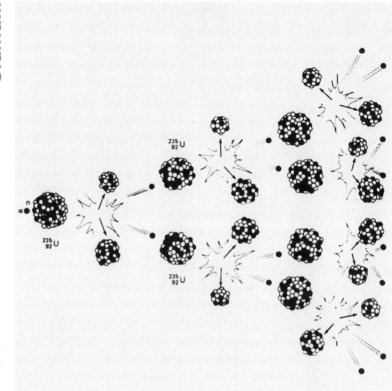

A nuclear chain reaction: the uninterrupted fissioning of ever-increasing numbers of uranium-235 atoms.

dant is a material that helps a dye stick to cloth. Uranium oxide has also found limited application as an attachment to filaments in lightbulbs. The compound reduces the speed at which an electric current enters the bulb. This reduces the likelihood of the filament heating too fast and breaking.

None of these applications is of very much importance today, however. By far the most important application is in nuclear weapons and nuclear power plants. The reason for this importance is that one isotope of uranium, uranium-235, undergoes nuclear fission.

Nuclear fission is the process by which neutrons are fired at a target. The target is usually made of uranium atoms. When neutrons hit the target, they cause the nuclei of uranium atoms to break apart. Smaller elements are formed and very large amounts of energy are given off.

When this reaction is carried out with no attempt to capture or control the energy, an enormous explosion takes place. This

Separating twins to make energy

Suppose neutrons are fired at a big block of uranium metal. Would nuclear fission occur? Would this be the way to make an atomic bomb? Could this process be used in a nuclear power plant?

The answer to all these questions is no. Only one isotope of uranium undergoes nuclear fission, uranium-235. The most common isotope, uranium-238, does not undergo fission. There is no way to make a bomb or a nuclear power plant with a chunk of natural uranium metal.

It is necessary is to increase the percentage of uranium-235 in the metal. As a chunk of uranium metal contains more uranium-235, it is more likely to undergo nuclear fission.

In making a bomb or a power plant, then, the first step is to separate the isotopes of uranium from each other. The goal is to produce more uranium-235 and less uranium-238.

That goal sounds easy, but it is very difficult to do. All isotopes of uranium behave very much alike. They have the same chemical properties. The only way they differ from each other is by weight. An atom of uranium-238, for example, weighs about 1 percent more than an atom of uranium-235. That's not much of a difference.

Scientists separate these isotopes in a centrifuge. A centrifuge is a machine that spins containers of materials at very high speeds. They are like some of the rides at an amusement park. A person sits in a compartment at the end of a long arm. When the ride is turned on, the compartment spins around faster and faster.

In a centrifuge, heavier objects spin farther out than do lighter objects. A mixture of uranium-235 and uranium-238 can be separated slightly in a centrifuge. But the separation is not very good because the isotopes weigh almost the same amount.

In practice, a mixture of isotopes must be centrifuged many times. Each time, the separation gets better.

Scientists prepare enriched uranium by this method. Enriched uranium contains more uranium-235 and less uranium-238. Enriched uranium was used to make atomic bombs and is now used in nuclear power plants. It contains enough uranium-235 to allow nuclear fission to occur.

release of nuclear energy accounts for the power of a nuclear weapon such as an atomic bomb. In reactors, the energy released during fission is used to boil water. Steam is produced and is converted to electricity. The controlled release of nuclear energy takes place in a nuclear power plant.

Today, over 400 nuclear power plants exist worldwide, producing about 17 percent of all electricity. Many people believe nuclear power will be more important in the future as the world's supply of coal, oil, and natural gas eventually runs out.

Other people are concerned about the dangers of nuclear power. The radiation released and radioactive wastes produced by nuclear power plants have made them unpopular in the United States. No new nuclear generators have been built for over ten years. It is not clear what the future of nuclear power plants in the United States will be.

Health effects

Since it is a radioactive element, uranium must be handled with great care. In addition, it poses an exceptional risk in powdered form. In this form, it tends to catch fire spontaneously.

Uranium poses an exceptional risk in powdered form. In this form, it tends to catch fire spontaneously.

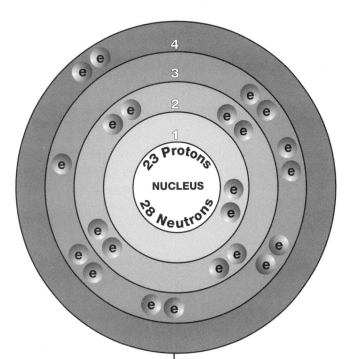

23 Protons

NUCLEUS

28 Neutrons

VANADIUM

Overview

Vanadium is a transition metal that lies toward the middle of the periodic table. The periodic table is a chart that shows how chemical elements are related to one another. Groups 4 through 12 are the transition metals.

Vanadium was discovered in 1801 by Spanish-Mexican metallurgist Andrés Manuel del Río (1764–1849). The element was re-discovered nearly 30 years later by Swedish chemist Nils Gabriel Sefström (1787–1845).

By far the most important application of vanadium today is in making alloys. An alloy is made by melting and mixing two or more metals. The mixture has properties different from those of the individual metals. Vanadium steel, for example, is more resistant to wear than ordinary steel. A potentially important new use of vanadium is in the manufacture of batteries. These batteries show promise for use in electric cars.

Discovery and naming

Andrés Manuel del Río was educated in France, Germany, and England, but moved to Mexico in 1794. There he became professor of mineralogy at the School of Mines in Mexico City.

SYMBOL
V

ATOMIC NUMBER
23

ATOMIC MASS
50.9415

FAMILY
Group 5 (VB)
Transition metal

PRONUNCIATION
vuh-NAY-dee-um

WORDS TO KNOW

Alloy a mixture of two or more metals with properties different from those of the individual metals

Battery a device for changing chemical energy into electrical energy

Catalyst a substance used to speed up or slow down a chemical reaction without undergoing any change itself

Ductile capable of being drawn into thin wires

Fly ash the powdery material produced during the production of iron or some other metal

Half life the time it takes for half of a sample of a radioactive element to break down

Isotopes two or more forms of an element that differ from each other according to their mass number

Metal elements that have a shiny surface, are good conductors of heat and electricity, can be melted, hammered into thin sheets, and drawn into thin wires

Non-metal elements that do not have the properties of metals

Slag a mixture of materials that separates from a metal during its purification and floats on top of the molten metal

While studying minerals at the School of Mines, he believed he had found a new element. He announced this discovery in 1801 and suggested the name panchromium, meaning "all colors." The new element formed compounds of many beautiful colors. Del Río later changed his mind and decided to call the element erythronium. The prefix *erytho-* means "red."

Del Río sent the mineral he was studying to colleagues in Europe for confirmation of his discovery. Unfortunately, they concluded that del Río's "new element" was **chromium.** Del Río's became discouraged and gave up his claim to the new element.

About thirty years later, however, del Río's element was discovered again. This time, the element was found by Sefström, who found the element in iron ore taken from a Swedish mine. He soon realized that his discovery was identical to that of del Río's. Vanadium was eventually named for the Scandinavian goddess of love, Vanadis.

Both Sefström and del Río saw vanadium only in the form of a compound, vanadium pentoxide (V_2O_5). It is very difficult to separate pure vanadium metal from this compound. It was not until 1887 that pure vanadium metal was isolated. English chemist Sir Henry Enfield Roscoe (1833–1915) found a way to separate pure vanadium from its oxide.

Physical properties

Vanadium is a silvery-white, ductile, metallic-looking solid. Ductile means capable of being drawn into thin wires. Its melting point is about 1,900°C (3,500°F) and its boiling point is about 3,000°C (5,400°F). Its density is 6.11 grams per cubic centimeter.

Chemical properties

Vanadium is moderately reactive. It does not react with **oxygen** in the air at room temperatures, nor does it dissolve in water. It does not react with some acids, such as hydrochloric or cold sulfuric acid. But it does become more reactive with hot acids, such as hot sulfuric and nitric acids.

Vanadium is special in that it acts like a metal in some cases, and as a non-metal in other cases. Metals are defined as elements that have a shiny surface, are good conductors of heat

Vanadium samples.

and electricity, can be melted, hammered into thin sheets, and drawn into thin wires. Non-metals generally do not have these properties.

Occurrence in nature

Vanadium is a relatively abundant element, ranking about 20th among elements occurring in the Earth's crust. Its abundance has been estimated at about 100 parts per million. That makes it about as abundant as **chlorine,** chromium, and **nickel.**

Vanadium is found in a number of minerals, including vanadinite, carnotite, roscoelite, and patronite. Commercially, however, it is obtained as a by-product of the manufacture of **iron.** Slag and fly ash are purified to remove the vanadium metal contained within them. Slag is a mixture of materials that separates from iron and floats on top of the molten metal. Fly ash is a powdery material produced during the purification of iron.

Vanadium is named after the Scandinavian goddess of love, Vanadis.

The vanadium obtained from slag and fly ash is usually in the form of ferrovanadium. Ferrovanadium is a mixture of iron and vanadium. It can be used in place of pure vanadium in making alloys. Ferrovanadium saves companies the cost of making pure vanadium metal.

Isotopes

Two naturally occurring isotopes of vanadium exist, vanadium-50 and vanadium-51. Vanadium-51 is much more common, making up about 99.75 percent of all naturally occurring vanadium.

Isotopes are two or more forms of an element. Isotopes differ from each other according to their mass number. The number written to the right of the element's name is the mass number. The mass number represents the number of protons plus neutrons in the nucleus of an atom of the element. The number of protons determines the element, but the number of neutrons in the atom of any one element can vary. Each variation is an isotope.

Vanadium-50 is radioactive. It has a half life of about 6 quadrillion years. The half life of a radioactive element is the time it takes for half of a sample of the element to break down. Ten grams of vanadium-50 today would reduce by 5 grams after 6 quadrillion years. The other half would have broken down to form a new isotope.

Extraction

Vanadium can be obtained in a variety of ways. For example, it can be produced by passing an electric current through molten (melted) vanadium chloride:

$$VCl_2 \xrightarrow{\text{electric current}} V + Cl_2$$

It can also be made by combining calcium metal with vanadium oxide:

$$5Ca + V_2O_5 \rightarrow 2V + 5CaO$$

Uses

About 90 percent of vanadium goes into steel alloys. Steel containing vanadium is stronger, tougher, and more rust-resistant than steel without vanadium. An important application of

such alloys is in space vehicles and aircraft. A common use of vanadium steel alloys is in tools used for cutting and grinding. Overall, about a third of all vanadium steel goes to building and heavy construction uses; less goes to transportation, machinery, and tool applications.

A knife with a vanadium steel alloy.

Compounds

The most important compound of vanadium commercially is vanadium pentoxide. Among its applications are as a catalyst for many industrial reactions, as a coloring material for glass and ceramics, and in the dyeing of textiles. A catalyst is a substance used to speed up or slow down a chemical reaction without undergoing any change itself.

An important new use for vanadium pentoxide may be in batteries. Scientists have been working for a very long time to make better batteries. Common automobile batteries are large and heavy. Batteries like these are too big and heavy for many applications. For example, they cannot be used in space probes and space vehicles. They weigh too much.

These batteries are also too large for use in electric cars. An electric car is powered by electricity rather than gasoline. A

great deal of research is being done toward the development of an economical electric car.

A new vanadium pentoxide battery produces more electrical energy per pound than the lead storage batteries in cars today. They are also likely to cause fewer environmental disposal problems.

Some manufacturers think electric cars with vanadium pentoxide batteries are part of the world's transportation future. Drivers would bring these cars into a "battery filling station." The worn-out battery would be "pumped out" and replaced in a matter of minutes. Economists predict that drivers may spend as much as $6 billion per year for such batteries by 2004!

Health effects

Vanadium occurs in living organisms in very small amounts. The total amount of vanadium in the human body is estimated to be less than 1 milligram (0.000035 ounce). It is found most commonly in the kidneys, spleen, lungs, testes, and bones. No specific function for vanadium in humans has been found.

Diseases due to a lack of vanadium have been found in rats, chicks, and goats, but only under artificial conditions produced by researchers. A lack of vanadium has never been found to have any health effects on any kind of animals in natural settings.

In large doses, vanadium can be toxic to humans and other animals. Its effects are not very serious, however. In general, the element and its compounds are not considered to be a hazard to health.

> Some manufacturers think electric cars with vanadium pentoxide batteries are part of the world's transportation future.

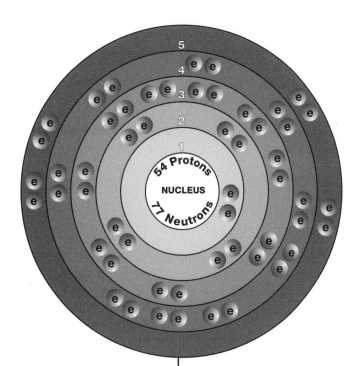

54 Protons

NUCLEUS

77 Neutrons

XENON

Overview

Xenon is a noble gas. The term noble gas is used to describe the elements in Group 18 (VIIIA) of the periodic table. The periodic table is a chart that shows how chemical elements are related to one another. "Noble gas" suggests a group of elements that is "too far above other elements" to react with them. The noble gases are also called the inert gases. That term has the same meaning. The noble gases only react with other elements under very unusual circumstances.

Xenon is very rare in the atmosphere. Its abundance is estimated to be about 0.1 parts per million. Xenon does not have many practical applications. Mostly, it is used to fill specialized lamps.

Discovery and naming

It took chemists more than a hundred years of careful research to understand the composition of air. In the early 1700s, they did not even understand the difference between the air around us and gases, like **oxygen,** carbon dioxide, and **nitrogen.** They used the word "air" to mean the same thing as "gas." Gases were very difficult to study. So it took a long time to figure out how various "airs" and "gases" differed from each other.

SYMBOL
Xe

ATOMIC NUMBER
54

ATOMIC MASS
131.29

FAMILY
Group 18 (VIIIA)
Noble gas

PRONUNCIATION
ZEE-non

Slowly the differences became apparent. In 1774, English chemist Joseph Priestley (1733–1804) realized he could remove a separate gas—oxygen—from air. Later, other gases in the air were identified. These included nitrogen, carbon dioxide, and other noble gases. One of the last gases to be isolated was xenon.

Xenon was discovered in 1898 by Scottish chemist and physicist Sir William Ramsay (1852–1916) and English chemist Morris William Travers (1872–1961). Ramsay and Travers used liquid air to make their discovery. Here's how they did the research:

If air is cooled to a very low temperature, it changes from a gas to a liquid. As it warms up, it changes back to a gas. But this change does not take place all at once. As liquid air warms, one gas (nitrogen) boils away first. As the temperature increases further, another gas (**argon**) boils off. Still later, a third gas (oxygen) boils off.

Great care must be used in doing this experiment. The first three gases to boil away (nitrogen, oxygen, and argon) make up 99.95 percent of air. It may look as if all the air is gone after the oxygen boils away, but it isn't.

After the oxygen is gone, a tiny bit of liquid air remains. That liquid air contains other atmospheric gases. One of those gases is xenon. Ramsay and Travers first recognized the presence of xenon in liquid air on July 12, 1898. They named the element xenon for the Greek word that means "stranger."

Physical properties

Xenon is a colorless, odorless gas. It has a boiling point of −108.13°C (−162.5°F) and a melting point of −111.80°C (−169.2°F). It may seem strange to talk about the "melting point" and "boiling point" of a gas. So think about the opposite of those two terms. The opposite of melting is "turning from a liquid into a solid." The opposite of boiling is "turning from a gas into a liquid."

Thus, the boiling point of xenon is the temperature at which the gas turns into a liquid. The melting point of xenon is the temperature at which liquid xenon turns into a solid.

WORDS TO KNOW

Inactive does not react with any other element

Inert gas *see* **Noble gas**

Isotopes two or more forms of an element that differ from each other according to their mass number

Liquid air air that has been cooled until all the gases in the air have condensed into a liquid

Noble gas an element in Group 18 (VIIIA) of the periodic table

Periodic table a chart that shows how chemical elements are related to one another

Radioactive isotope an isotope that breaks apart and gives off some form of radiation

William Ramsay.

The density of xenon gas is 5.8971 grams per liter. That makes xenon about four times as dense as air.

Chemical properties

For many years, xenon was thought to be completely inactive. Inactive means that it does not react with any other element. Then, in 1962, English chemist Neil Bartlett (1932–) made xenon platinofluoride ($XePtF_6$). Bartlett's success inspired other chemists to try making other xenon compounds. Chemists found ways to make such xenon compounds as xenon difluoride (XeF_2), xenon tetrafluoride (XeF_4), xenon hexafluoride (XeF_6), xenon trioxide (XeO_3), and xenon oxytetrafluoride ($XeOF_4$).

Occurrence in nature

The Earth's atmosphere contains about 0.1 part per million of xenon. Studies indicate that the atmosphere of Mars may contain about the same amount of xenon, perhaps 0.08 parts per million. The element is not known to occur in the Earth's crust.

In the early 1700s, they did not even understand the difference between the air around us and gases, like oxygen, carbon dioxide, and nitrogen. They used the word "air" to mean the same thing as "gas."

Lead canisters used to store radioactive xenon for medical diagnostic purposes.

Isotopes

Nine naturally occurring isotopes of xenon exist. They are xenon-124, xenon-126, xenon-128, xenon-129, xenon-130, xenon-131, xenon-132, xenon-134, and xenon-136. Isotopes are two or more forms of an element. Isotopes differ from each other according to their mass number. The number written to the right of the element's name is the mass number. The mass number represents the number of protons plus neutrons in the nucleus of an atom of the element. The number of protons determines the element, but the number of neutrons in the atom of any one element can vary. Each variation is an isotope.

At least 18 radioactive isotopes of xenon are known also. A radioactive isotope is one that breaks apart and gives off some form of radiation. Radioactive isotopes are produced when very small particles are fired at atoms. These particles stick in the atoms and make them radioactive.

Two radioactive isotopes of xenon—xenon-127 and xenon-133—are used in medicine. These isotopes are used to study the flow of blood through the brain and the flow of air through the lungs. In most cases, the patient inhales the radioactive gas through a mask. The xenon gas moves through the body just like oxygen or any other gas. As it travels through the body, the xenon isotope gives off radiation. The radiation can be detected by measuring devices held over the body. Doctors can tell whether the patient's lungs are working properly.

Extraction

Xenon is produced in the same way it was discovered. Liquid air is allowed to evaporate. When most other gases have boiled off, xenon is left behind. The techniques used today are much better than those used by Ramsay and Travers, of course. It is now relatively easy to capture the xenon gas in air by this method.

Uses

The primary use of xenon is in lamps. When an electric current is passed through a gas, it can give off light. Fluorescent lamps and "neon" lights are examples of this process. The kind and color of light given off depend on the gas used in the lamp. Xenon is used when a very bright, sun-like light is needed. For example, the flash units and bright lights used by photographers are often made with xenon gas.

Ultraviolet lights used to sterilize laboratory equipment may also contain xenon. The light produced is strong enough to kill bacteria. Xenon is also used in the manufacture of strobe lights. A strobe light produces a very bright, intense light in very short pulses. Strobe lights appear to "freeze" the movement of an object. Each time the light flashes on, it shines on the moving object for a fraction of a second. The object's motion can be broken down into any number of very short intervals.

Compounds

So far, xenon compounds are only laboratory curiosities. They have no practical applications. (See under "Chemical properties.")

Health effects

Xenon is a harmless gas. Some of its compounds, however, are toxic.

Two radioactive isotopes of xenon are used to study the flow of blood through the brain and the flow of air through the lungs.

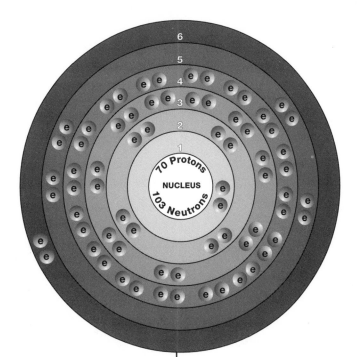

YTTERBIUM

Overview

Ytterbium belongs to the lanthanide family. The lanthanides make up Row 6 of the periodic table. The periodic table is a chart that shows how the chemical elements are related to each other. The lanthanides are also known as the rare earth elements. The name suggests that the lanthanides do not occur commonly in the earth. In fact, that is not correct. They are not all that uncommon. The name rare earth arose because the elements are so difficult to separate from each other. With modern techniques, this separation can be done much more easily.

Ytterbium was one of nine new elements discovered in the mineral yttria at the end of the nineteenth century. Analyzing this mineral posed great difficulties for chemists of the time. The elements in yttria have very similar properties. That makes it difficult to separate them from each other. Three chemists, Jean-Charles Gallisard de Marignac, Lars Fredrik Nilson, and Georges Urbain, all deserve partial credit for discovering ytterbium.

Discovery and naming

In 1878, French chemist Jean-Charles-Galissard de Marignac (1817–94) reported his analysis of the mineral erbia. Erbia was one of the minerals found a century earlier in an interesting new

SYMBOL
Yb

ATOMIC NUMBER
70

ATOMIC MASS
173.04

FAMILY
Lanthanide
(rare earth metal)

PRONUNCIATION
i-TER-bee-um

rock called yttria. The rock had been discovered outside the town of Ytterby, Sweden, by Swedish army officer Carl Axel Arrhenius (1757–1824) in 1787. In the century that followed Arrhenius' discovery, chemists worked hard to find out what elements were in yttria. Earlier chemists thought erbia was a new element, but Marignac disagreed. He said that erbia consisted of two new elements, which he called **erbium** and ytterbium.

The very next year, a second Swedish chemist, Lars Fredrik Nilson (1840–99), proved that Marignac was wrong. Ytterbium was not a new element, he said. Instead, it consisted of two other new elements. Nilson called these elements scandium and ytterbium. (See sidebar on Nilson in the **scandium** entry.)

Nilson's analysis still did not solve this confusion. In 1907, French chemist Georges Urbain (1872–1938) announced that Nilson's ytterbium was also a mixture of two new elements. Urbain called these elements ytterbium and **lutetium.** Marignac, Nilson, and Urbain are all given part of the credit for the discovery of ytterbium.

In fact, the ytterbium studied by Marignac, Nilson, and Urbain was not pure ytterbium. Instead, it was combined with **oxygen** and other elements. Fairly pure ytterbium metal was not produced until 1937 and high purity ytterbium was not produced until 1953.

Physical properties
Ytterbium is a typical metal. It has a bright, shiny surface and is malleable and ductile. Malleable means capable of being hammered into thin sheets. Ductile means capable of being drawn into thin wires. Its melting point is 824°C (1,515°F) and its boiling point is 1,427°C (2,600°F). It has a density of 7.01 grams per cubic centimeter.

Chemical properties
Ytterbium tends to be more reactive than other lanthanide elements. It is usually stored in sealed containers to keep it from reacting with oxygen in the air. It also reacts slowly with water and more rapidly with acids and liquid ammonia.

Occurrence in nature
Ytterbium is one of the more common lanthanides. It is thought to have an abundance of about 2.7 to 8 parts per mil-

WORDS TO KNOW

Ductile capable of being drawn into thin wires

Isotopes two or more forms of an element that differ from each other according to their mass number

Lanthanides the elements in the periodic table with atomic numbers 58 through 71

Laser a device for making very intense light of one very specific color that is intensified many times over

Malleable capable of being hammered into thin sheets

Radioactive isotope an isotope that breaks apart and gives off some form of radiation

Rare earth elements *see* **Lanthanides**

lion in the Earth's crust. That makes it somewhat more common than **bromine, uranium, tin,** and **arsenic.** Its most common ore is monazite, which is found in beach sands in Brazil, India, and Florida. Monazite typically contains about 0.03 percent ytterbium.

Isotopes

Seven naturally occurring isotopes of ytterbium are known. These isotopes are ytterbium-168, ytterbium-170, ytterbium-171, ytterbium-172, ytterbium-173, ytterbium-174, and ytterbium-176. Isotopes are two or more forms of an element. Isotopes differ from each other according to their mass number. The number written to the right of the element's name is the mass number. The mass number represents the number of protons plus neutrons in the nucleus of an atom of the element. The number of protons determines the element, but the number of neutrons in the atom of any one element can vary. Each variation is an isotope.

Ten radioactive isotopes of ytterbium are known also. A radioactive isotope is one that breaks apart and gives off some form of radiation. Radioactive isotopes are produced when very small particles are fired at atoms. These particles stick in the atoms and make them radioactive.

Studies have been done on one radioactive isotope of ytterbium, ytterbium-169, for possible use in a portable X-ray machine. This isotope gives off gamma radiation, which is similar to X rays. Gamma rays pass through soft tissues in the body, just like X rays. But they are blocked by bones and other thick material. A small amount of ytterbium-169 acts just like a tiny X-ray machine. It can be carried around more easily than can a big X-ray machine.

Extraction

Ytterbium is obtained from its ores by reaction with **lanthanum** metal:

$$2La + Yb_2O_3 \rightarrow 2Yb + La_2O_3$$

Uses

Ytterbium has no major commercial uses. A small amount is used to add strength to special types of steel. Some ytterbium is also used in making lasers. A laser is a device for producing

Ytterbium's most common ore is found in beach sands in Brazil, India, and Florida.

very bright light of a single color. The kind of light produced by a laser depends on the elements used in making it. A laser made with ytterbium has properties different from those without ytterbium.

Compounds

The only ytterbium compound of commercial interest is ytterbium oxide (Yb_2O_3). This compound is used to make alloys and special types of ceramics and glass.

Health effects

Ytterbium is not thought to be a very toxic element.

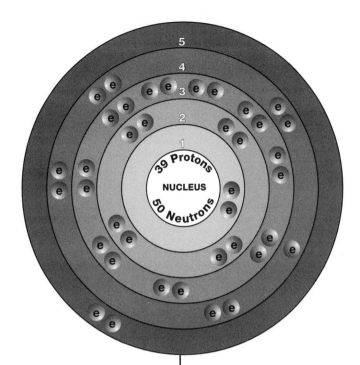

YTTRIUM

Overview

Yttrium is one of four elements named for the same small town of Ytterby, Sweden. The other three elements are **erbium, terbium,** and **ytterbium.** The element was discovered in 1794 by Finnish chemist Johan Gadolin (1760–1852). The discovery of yttrium marked the beginning of one hundred years of complicated chemical research that resulted in the discovery of ten new elements.

Yttrium is a transition metal. Transition metals are those elements in Groups 3 through 12 of the periodic table. The periodic table is a chart that shows how chemical elements are related to each other. The element above yttrium in the periodic table is **scandium.** The space below yttrium is taken up by a group of elements known as the rare earth elements. Scandium, yttrium, and the rare earth elements are often found together in nature.

Yttrium is often used to make alloys with other metals. An alloy is made by melting and mixing two or more metals. The mixture has properties different from those of the individual metals. Two of yttrium's most interesting applications are in lasers and superconducting materials.

SYMBOL
Y

ATOMIC NUMBER
39

ATOMIC MASS
88.9059

FAMILY
Group 3 (IIIB)
Transition metal

PRONUNCIATION
I-tree-um

Alloy a mixture of two or more metals with properties different from those of the individual metals

"Doped" containing a small amount of a material as an impurity

Isotopes two or more forms of an element that differ from each other according to their mass number

Lanthanides the elements in the periodic table with atomic numbers 58 through 71

Laser a device for making very intense light of one very specific color that is intensified many times over

Periodic table a chart that shows how chemical elements are related to each other

Phosphor a material that gives off light when struck by electrons

Radioactive isotope an isotope that breaks apart and gives off some form of radiation

Rare earth elements *see* Lanthanides

Superconductor a material that has no resistance to the flow of electricity; once an electrical current begins flowing in the material, it continues to flow forever

Transition metal an element in Groups 3 through 12 of the periodic table

A laser is a device for producing very bright light of a single color. One of the most popular lasers is made of yttrium, **aluminum,** and garnet. Garnet is a gem-like material with a sand-like composition. Superconducting materials are substances with no resistance to the flow of an electric current. An electric current that begins to flow through them never stops. Superconducting materials may have many very important applications in the future.

Discovery and naming

In 1787, a lieutenant in the Swedish army named Carl Axel Arrhenius (1757–1824) found an interesting new stone near Ytterby. He gave the stone to Gadolin for analysis. At the time, Gadolin was professor of chemistry at the University of Åbo in Finland. Gadolin decided that Arrhenius' rock contained a new element. That element was later given the name yttrium.

For about fifty years, nothing new was learned about yttrium. Then Swedish chemist Carl Gustav Mosander (1797–1858) discovered that yttrium was not a single pure substance. Instead, it was a mixture of three new substances. In addition to Gadolin's yttrium, Mosander found two more elements. He called these elements terbium and erbium.

From that point on; the story of yttrium continued to get more and more complicated. As it turned out, neither terbium nor erbium was a pure element. Both new "elements" also contained other new elements. And these new "elements," in turn, contained other new elements. In the end, the heavy black mineral found by Arrhenius resulted in the discovery of ten new elements! (See the individual entries for the nine other elements: **dysprosium, erbium, gadolinium, holmium, lutetium, scandium, terbium, thulium,** and **ytterbium.**)

Physical properties

Yttrium has a bright, silvery surface, like most other metals. It is also prepared as a dark gray to black powder with little shine. Yttrium has a melting point of 1,509°C (2,748°F) and a boiling point of about 3,000°C (5,400°F). Its density is 4.47 grams per cubic centimeter.

Chemical properties

The chemical properties of yttrium are similar to those of the rare earth elements. It reacts with cold water slowly, and with

Yttrium is sometimes included in superconducting materials. Here, a small magnet hovers over a nitrogen-cooled specimen of a superconducting ceramic.

hot water very rapidly. It dissolves in both acids and alkalis. An alkali is the chemical opposite of an acid. Sodium hydroxide ("household lye") and limewater are common alkalis.

Solid yttrium metal does not react with **oxygen** in the air. However, it reacts very rapidly when in its powdered form. Yttrium powder may react explosively with oxygen at high temperatures.

as abundant as **cobalt, copper,** and **zinc.** As with other elements, the abundance of yttrium is quite different in other parts of the solar system. Rocks brought back from the Moon, for example, have a high yttrium content.

Yttrium occurs in most rare earth minerals. A rare earth mineral contains one or more—usually many—of the rare earth elements. The most important rare earth mineral is monazite. Monazite occurs in many places in the world, especially Brazil,

Making machines work more efficiently

One of the important new uses for yttrium is in superconductors. Superconductors were first discovered by Dutch physicist Heike Kamerlingh-Onnes (1853–1926) in 1911. Kamerlingh-Onnes found that certain metals cooled to nearly absolute zero lost all resistance to an electric current. Absolute zero is the coldest temperature possible, about –273°C. Once an electric current got started in these very cold metals, it could keep going forever. These metals were called superconductors.

Research on superconductors did not advance very much for seventy years. It is very difficult to produce temperatures close to absolute zero, and it is difficult to work with materials at these temperatures.

Then, in 1986, a startling announcement was made. Two scientists at the IBM Research Laboratories in Zürich, Switzerland, had made a material that becomes superconducting at 35 degrees above absolute zero. That temperature, –238°C, is still very cold, but it is much "warmer" than the temperature at which Kamerlingh-Onnes had worked.

An even bigger jump was announced only a year later. A team of researchers working under Ching-Wu "Paul" Chu (1941–) produced superconductors that worked at 90 to 100 degrees above absolute zero. These temperatures are also still very cold, but they broke an important barrier. Those temperatures are close to the temperature of liquid nitrogen. Scientists have known how to make and work with liquid nitrogen for several hundred years. It had now become easy to work with superconducting materials.

These "high-temperature" superconducting materials are very interesting. In the first place, they are not metals. They are ceramics. A ceramic is a clay-like material. It often consists of sand, clay, brick, glass or a stone-like material.

In the second place, the composition of these materials is difficult to determine. They usually contain barium, copper, lanthanum, yttrium, and oxygen. They often contain other elements. But they are not simple compounds, like copper oxide (CuO) or yttrium oxide (Y2O3). Instead, they are complex mixtures of the elements.

Superconductors may be very important materials in the future. Electrical machinery usually does not operate very efficiently. The electric current has to work hard to overcome resistance in wires and other parts of the machinery. A lot of the electrical energy is lost because of this resistance. The electrical energy turns into heat.

In a machine made of superconducting materials, the electrical current would meet no resistance at all. All of the electrical energy could be used productively. It could make the machine operate, rather than being lost as heat.

Australia, Canada, and parts of the United States. Typically, monazite contains about 3 percent yttrium.

Isotopes
There is only one naturally occurring isotope of yttrium, yttrium-89. Isotopes are two or more forms of an element. Iso-

topes differ from each other according to their mass number. The number written to the right of the element's name is the mass number. The mass number represents the number of protons plus neutrons in the nucleus of an atom of the element. The number of protons determines the element, but the number of neutrons in the atom of any one element can vary. Each variation is an isotope.

commercial use. However, yttrium-90 is now being tested as a treatment for cancer. Radiation given off by the isotope kills cancer cells. Researchers believe that yttrium-90 may find wider use in the future for treating cancer. One advantage of using this isotope is that is easy to obtain. It is produced when another radioactive isotope (strontium-90) breaks down. Strontium-90 is a by-product formed in nuclear power plants.

Extraction

Yttrium is usually bought and sold in the form of yttrium oxide (Y_2O_3). However, the pure metal can be obtained by combining another compound of yttrium, yttrium fluoride (YF_3), with calcium metal at high temperatures:

$$2YF_3 + 3Ca \rightarrow 3CaF_2 + 2Y$$

Uses

Traditionally, yttrium has had many of the same uses as the rare earth elements. For example, it has been used in phosphors. A phosphor is a material that shines when struck by electrons. The color of the phosphor depends on the elements of which it is made. Yttrium phosphors have long been used in color television sets and in computer monitors. They have also been used in specialized fluorescent lights. In 1996, about two-thirds of all the yttrium consumed was used for these purposes.

Yttrium alloys have some special uses as well. These alloys tend to be hard, resistant to wear, and resistant to corrosion (rusting). They are used in cutting tools, seals, bearings, and jet engine coatings. Slightly less than a third of all yttrium used in 1996 went to applications like these.

One of the areas in which yttrium is becoming more important is in the manufacture of lasers. Lasers are devices for producing very intense beams of light of a single color. These beams are used for precision metal cutting and surgery. There is some hope that lasers may someday replace the dental drill.

One of the most widely used lasers today is the yttrium-aluminum-garnet (YAG) laser. YAG lasers often contain other elements. These elements change the kind of light produced by the laser in one way or another. The laser is said to be doped with another element if it contains a small amount of that element. An example of this kind of laser is one doped with **neodymium.** The neodymium-doped YAG (Nd:YAG) laser has been used to make long distance measurements.

In this kind of laser, a beam is fired at a far-away object. The time it takes for the beam to be reflected is then measured. The time is used to calculate the distance to the distant object. One application of this principle is used by space probes. For example, on February 17, 1996, the National Aero-

> There is some hope that lasers may someday replace the dental drill.

nautics and Space Administration (NASA) launched a space-craft to observe the asteroid Eros. The back side of the asteroid will be measured with a Nd:YAG laser. Beams from the laser will be used to map surface features on the asteroid.

Compounds

The only yttrium compound of commercial interest is yttrium oxide (Y_2O_3). Yttrium oxide is used to make phosphors for color television sets and in crystals used in microwave detection instruments.

Health effects

Yttrium has been found to be toxic to laboratory rats in high doses. However, there is little information about its effects on humans. In such cases, an element is usually treated as if it were dangerous.

The back side of the asteroid Eros will be measured with a neodymium-doped yttrium-aluminum-garnet (Nd:YAG) laser.

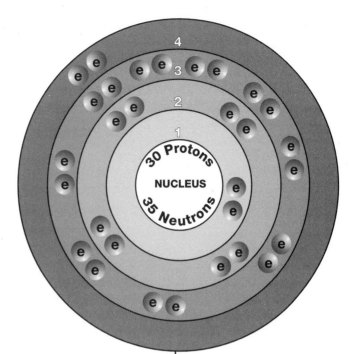

ZINC

Overview

Zinc is a transition metal that occurs in the center of the periodic table. The periodic table is a chart that shows how chemical elements are related to each other. The space between Groups 2 and 13 is occupied by the transition metals. These metals share many physical and chemical properties in common.

Alloys and compounds of zinc have been known since at least 500 B.C. But zinc metal was not known or used until much later. The reason is that zinc boils away or vaporizes easily when heated. Any effort to release zinc from its compounds also causes the metal to evaporate into the air.

Zinc was probably known in Asia before it was discovered in Europe. Ancient books from both India and China refer to zinc products. Such products were imported to Europe from Asia before they were made in Europe.

The most important use of zinc today is in galvanizing other metals. Galvanizing is the process of laying down a thin layer of zinc on the surface of a second metal. Zinc does not corrode (rust) as easily as **iron** and other metals. So the thin layer of zinc protects iron and other metals from corrosion.

SYMBOL
Zn

ATOMIC NUMBER
30

ATOMIC MASS
65.38

FAMILY
Group 12 (IIB)
Transition metal

PRONUNCIATION
ZINK

Alchemy a kind of pre-science that existed from about 500 B.C. to about the end of the 16th century

Alloy a mixture of two or more metals with properties different from those of the individual metals

Ductile capable of being drawn into thin wires

Galvanizing the process of laying down a thin layer of zinc on the surface of a second metal

Isotopes two or more forms of an element that differ from each other according to their mass number

Malleable capable of being hammered into thin sheets

Periodic table a chart that shows how chemical elements are related to each other

Radioactive isotope an isotope that breaks apart and gives off some form of radiation

Sublimation the process by which a solid changes directly to a gas when heated, without first changing to a liquid

Toxic poisonous

Tracer a radioactive isotope whose presence in a system can easily be detected

Transition metal an element in Groups 3 through 12 of the periodic table

Discovery and naming

Some metals can be obtained from their ores easily. In a few cases, all that is needed is to heat the ore. Heating an ore of zinc releases the free metal. But with zinc, there is an additional problem. Zinc metal sublimates very easily. Sublimation is the process by which a solid changes directly to a gas when heated, without first changing to a liquid. Anyone who wanted to make zinc from its ore would lose the zinc almost immediately by sublimation.

Of course, early people did not understand this process. They may very well have made zinc by heating its ores. But any zinc they made would have floated away immediately. Still, a process for extracting zinc from its ores was apparently invented in India by the 13th century. The process involves heating the zinc ore in a closed container. When zinc vapor forms, it condenses inside the container. It can then be scraped off and used. That method seems to have been passed to China and then, later, to Europe.

In the meantime, ancient people were familiar with compounds and alloys of zinc. For example, there are brass objects from Palestine dating to 1300 B.C. Brass is an alloy of **copper** and zinc. The alloy may have been made by humans or found naturally in the earth. No one knows the origin of the brass in these objects.

The first European to describe zinc was probably Swiss physician Paracelsus (1493–1541). Paracelsus' real name was Theophrastus Bombastus von Hohenheim. Early in life, he took the name Paracelsus, meaning "greater than Celsus." Celsus was one of the great Roman physicians. Paracelsus wanted the world to know that he was even "greater than Celsus."

Paracelsus was also an alchemist. Alchemy was a kind of pre-science that existed from about 500 B.C. to near the end of the 16th century. People who studied alchemy—alchemists—wanted to find a way to change **lead, iron,** and other metals into **gold.** They were also looking for the "secret to eternal life." Alchemy contained too much magic and mysticism to be a real science. But it developed a number of techniques and produced many new materials that were later found to be useful in modern chemistry.

Paracelsus first wrote about zinc in the early 1500s. He described some properties of the metal. But he said he did not know what the metal was made of. Because of his report on the metal, Paracelsus is sometimes called the discoverer of zinc.

The name zinc was first used in 1651. It comes from the German name for the element, *Zink*. What meaning that word originally had is not known.

Physical properties
Zinc is a bluish-white metal with a shiny surface. It is neither ductile nor malleable at room temperature. Ductile means capable of being drawn into thin wires. Malleable means capable of being hammered into thin sheets. At temperatures above 100°C (212°F), however, zinc becomes somewhat malleable.

Zinc's melting point is 419.5°C (787.1°F) and its boiling point is 908°C (1,670°F). Its density is 7.14 grams per cubic centimeter. Zinc is a fairly soft metal. Its hardness is 2.5 on the Mohs scale. The Mohs scale is a way of expressing the hardness of a material. It runs from 0 (for talc) to 10 (for diamond).

A process for extracting zinc from its ores was apparently invented in India by the 13th century.

Chemical properties
Zinc is a fairly active element. It dissolves in both acids and alkalis. An alkali is a chemical with properties opposite those of an acid. Sodium hydroxide ("common lye") and limewater are examples of alkalis. Zinc does not react with oxygen in dry air. In moist air, however, it reacts to form zinc carbonate. The zinc carbonate forms a thin white crust on the surface which prevents further reaction. Zinc burns in air with a bluish flame.

Occurrence in nature
The abundance of zinc in the Earth's crust is estimated to be about 0.02 percent. That places the element about number 24 on the list of the elements in terms of their abundance.

Zinc never occurs as a free element in the earth. Some of its most important ores are smithsonite, or zinc spar or zinc carbonate ($ZnCO_3$); sphalerite, or zinc blende or zinc sulfide (ZnS); zincite, or zinc oxide (ZnO); willemite, or zinc silicate ($ZnSiO_3$); and franklinite [$(Zn,Mn,Fe)O \cdot (Fe,Mn_2)O_3$].

The largest producer of zinc ore in the world today is Canada. Other important producing nations are Australia, China, Peru, the United States, and Mexico. In the United States, more than half of the zinc produced comes from Alaska. Other important producing states are Tennessee, Missouri, Montana, and New York.

Isotopes

Five naturally occurring isotopes of zinc are known. They are zinc-64, zinc-66, zinc-67, zinc-68, and zinc-70. Isotopes are two or more forms of an element. Isotopes differ from each other according to their mass number. The number written to the right of the element's name is the mass number. The mass number represents the number of protons plus neutrons in the nucleus of an atom of the element. The number of protons determines the element, but the number of neutrons in the atom of any one element can vary. Each variation is an isotope.

About eight radioactive isotopes of zinc are known also. A radioactive isotope is one that breaks apart and gives off some form of radiation. Radioactive isotopes are produced when very small particles are fired at atoms. These particles stick in the atoms and make them radioactive.

One radioactive isotope of zinc, zinc-65, has some practical importance. Zinc-65 is used as a tracer to study physical and biological events. A tracer is an isotope whose presence in a system can easily be detected. The isotope is injected into the system at some point. Inside the system, the isotope gives off radiation. That radiation can be followed by means of detectors placed around the system.

For example, zinc-65 is used to study how alloys wear out. An alloy can be made using zinc metal. But the zinc used is zinc-65 instead of ordinary zinc. Changes in radiation given off by the radioactive isotope can be followed to find patterns in the way the alloy wears out. Zinc-65 can also be used to study the role of zinc in the human body. A person can be fed food that contains a small amount of zinc-65. The movement of the isotope through the body can be followed with a detector. A researcher can see where the isotope goes and what roles it plays in the body.

Brass is an alloy of **copper** and zinc.

Extraction

As with many metals, pure zinc can be prepared from an ore by one of two methods. First, the ore can be roasted (heated in air). Roasting converts the ore to a compound of zinc and oxygen, zinc oxide (ZnO). The compound can then be heated with charcoal (pure **carbon**). The carbon takes the oxygen away from the zinc, leaving the pure metal behind:

$$2ZnO + C \xrightarrow{\text{heated}} 2Zn + CO_2$$

The other method is to pass an electric current through a compound of zinc. The electric current causes the compound to break apart. Pure zinc metal is produced.

Uses

The annual cost of corrosion (rusting) in the United States is estimated to be about $300 billion. This is money lost when metals become corroded and break apart. Buildings and

Zinc oxide manufacturing. This compound is used to produce textiles, storage batteries, paints, and rubber products.

Zinc burns in air with a bluish flame.

bridges are weakened, cars and trucks rust, farm equipment breaks down, and metal used in many other applications is destroyed. It is hardly surprising that protecting metal from corrosion is an important objective in American industry. One of the most effective ways of providing protection is through galvanizing. Today, about half of all the zinc produced in the United States is used to galvanize other metals. The largest consumers of galvanized metal are the construction and automotive industries.

The second largest use of zinc is in making alloys. An alloy is made by melting and mixing two or more metals. The mixture has properties different from those of the individual metals. Two of the most common alloys of zinc are brass and bronze. Brass is an alloy of zinc and copper. Bronze is an alloy of copper and **tin** that may also contain a small amount of zinc. Alloys of zinc are used in a great variety of products, including automobile parts, roofing, gutters, batteries, organ pipes, electrical fuses, type metal, household utensils, and building materials.

Compounds

A number of zinc compounds have important uses. Some examples are the following:

zinc acetate ($Zn(C_2H_3O_2)_2$): wood preservative; dye for textiles; additive for animal feed; glazing for ceramics

zinc arsenate ($Zn_3(AsO_4)_2$): wood preservative; insecticide

zinc borate (ZnB_4O_7): fireproofing of textiles; prevents the growth of fungus and mildew

zinc chloride ($ZnCl_2$): solder (for welding metals); fireproofing; food preservative; additive in antiseptics and deodorants; treatment of textiles; adhesives; dental cement; petroleum refining; and embalming and taxidermy products

zinc fluorosilicate ($ZnSiF_6$): mothproofing agent; hardener for concrete

zinc hydrosulfite (ZnS_2O_4): bleaching agent for textiles, straw, vegetable oils, and other products;

Zinc deficiency can interfere with a plant's ability to reproduce.

Opposite page:
The pipes of this beautiful church organ are made of zinc alloys.

Zinc alloys are used in the production of electrical fuses.

brightening agent for paper and beet and cane sugar juice

zinc oxide (ZnO): used in rubber production; white pigment in paint; prevents growth of molds on paints; manufacturer of glass; photocopy machines; production of many kinds of glass, ceramics, tile, and plastics

zinc phosphide (Zn_3P_2): rodenticide (rat killer)

zinc sulfate ($ZnSO_4$): manufacture of rayon; supplement in animal feeds; dyeing of textiles; and wood preservative

Health effects

Zinc is an essential micronutrient for plants, humans, and animals. Zinc deficiency has relatively little effect on the health of a plant, but it interferes with reproduction. Pea plants deprived of zinc, for example, will form flowers. But the flowers will not turn to seeds.

In humans, zinc deficiencies are more serious. Zinc is used to build molecules of DNA. DNA is the chemical in our body that tells cells what chemicals they should make. It directs the reproduction of humans also. Fetuses (babies that have not yet been born) deprived of zinc may grow up to have mental or physical problems. Young children who do not get enough zinc in their diet may experience loss of hair and skin lesions. They may also experience retarded growth called dwarfism. Chemists have now found that zinc plays an essential role in the manufacture of many important chemicals in the human body.

On the other hand, an excess of zinc can cause health problems, too. Breathing zinc dust may cause dryness in the throat, coughing, general weakness and aching, chills, fever, nausea, and vomiting. One sign of zinc poisoning is a sweet taste in the mouth that cannot be associated with eating sweet foods. Certain compounds of zinc can be harmful to health also. Zinc chloride ($ZnCl_2$), for example, can cause skin rashes and sore throat.

Zinc is an essential micronutrient for humans. But too much or too little can cause health problems.

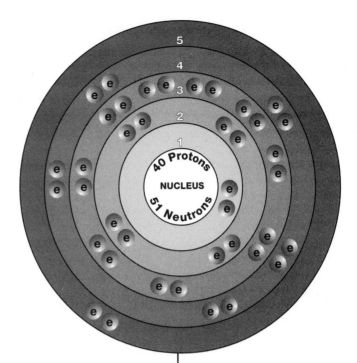

ZIRCONIUM

Overview

Compounds of zirconium have been known for centuries. Yet, the element itself was not recognized until 1789. In that year, German chemist Martin Heinrich Klaproth (1743–1817) discovered the element in a stone brought to him from the island of Ceylon (now Sri Lanka).

Zirconium is one of the transition metals. The transition metals are the elements found in Rows 4 through 7 and between Groups 2 and 13 in the periodic table. The periodic table is a chart that shows how chemical elements are related to each other. Zirconium is located below **titanium,** which it resembles, in the periodic table. Below zirconium is **hafnium,** a chemical twin of zirconium.

An important use of zirconium is in nuclear power plants. Its most important compound is zircon, which has a number of industrial applications. Zircon can also be obtained in gemstone quality. A gemstone is a mineral that can be cut and polished and used in jewelry or art.

Discovery and naming

Naturally occurring compounds of zirconium have been used by humans since before the birth of Christ. For example, St.

SYMBOL
Zr

ATOMIC NUMBER
40

ATOMIC MASS
91.22

FAMILY
Group 4 (IVB)
Transition metal

PRONUNCIATION
zir-KO-nee–um

John talks about the jacinth (or hyacinth) stone. He says it was one of the jewels found in the walls surrounding Jerusalem. The jacinth stone was the same mineral referred to by the Persians as *zargun,* meaning "gold-like" in Persian.

Early chemists did not study the jacinth stone very carefully. They thought it was another form of alumina (aluminum oxide). Alumina was a well-known mineral at the time. In fact, it was not until Klaproth undertook the study of the jacinth stone that he realized it contained a new element. Klaproth at first referred to the stone as Jargon of Ceylon. When he knew that he had found a new element, he suggested the name zirconium for it.

The material discovered by Klaproth was not a pure element. Instead, it was a compound of zirconium and **oxygen,** zirconium oxide (ZrO_2). The pure metal was not produced until 1824 when Swedish chemist Jöns Jakob Berzelius (1779–1848) made fairly pure zirconium. He made the metal by heating a mixture of **potassium** and potassium zirconium fluoride (ZrK_2F_6):

$$4K + ZrK_2F_6 \xrightarrow{\text{heated}} 6KF + Zr$$

Physical properties

Zirconium is a hard, grayish-white, shiny metal. Its surface often has a flaky-like appearance. It also occurs in the form of a black or bluish-black powder. It has a melting point of 1,857°C (3,375°F) and a boiling point of 3,577°C (6,471°F). Its density is 6.5 grams per cubic centimeter.

Zirconium has one physical property of special importance: It is transparent to neutrons. Neutrons are tiny particles with no charge in the nucleus (center) of almost all atoms. Industrially, they are used to make nuclear fission reactions occur. Nuclear fission is the process in which large atoms break apart. Large amounts of energy and smaller atoms are produced during fission. Fission reactions are used to provide the power behind nuclear weapons (such as the atomic bomb). They are also used to produce energy in a nuclear power plant.

One of the difficult problems in building a nuclear power plant is selecting the right materials. Many metals capture neutrons that pass through them. The neutrons become part of the metal atoms and are no longer available to make fission reac-

WORDS TO KNOW

Abrasive a powdery material used to grind or polish other materials

Alloy a mixture of two or more metals with properties different from those of the individual metals

Isotopes two or more forms of an element that differ from each other according to their mass number

Periodic table a chart that shows how chemical elements are related to each other

Radioactive isotope an isotope that breaks apart and gives off some form of radiation

Refractory a material that does not conduct heat well

Transition metal an element in Groups 3 through 12 of the periodic table

A zirconium sample.

tions occur. An engineer needs to use materials in a power plant that are transparent to neutrons—that is, that allow neutrons to pass through them.

Zirconium is one of the best of these metals. If zirconium is used to make the parts in a nuclear power plant, it will not remove neutrons from the fission reaction going on inside the plant.

A special alloy of zirconium has been developed just for this purpose. It is called Zircaloy. The manufacture of Zircaloy accounts for 90 percent of the zirconium metal used in the world today.

Chemical properties

Zirconium is a fairly inactive element. When exposed to air, it reacts with oxygen to form a thin film of zirconium oxide (ZrO_2). This film protects the metal from further corrosion (rusting). Zirconium does not react with most cold acids or with water. It does react with some acids that are very hot, however.

Occurrence in nature

Zirconium is a fairly common element in the Earth's crust. Its abundance is estimated to be 150 to 230 parts per million.

That places it just below **carbon** and **sulfur** among elements occurring in the Earth's crust.

The two most common ores of zirconium are zircon, or zirconium silicate ($ZrSiO_4$); and baddeleyite, or zirconia or zirconium oxide (ZrO_2). The amount of zirconium produced in the United States is not reported. That information is regarded as a trade secret. The largest suppliers of zirconium minerals in the world are Australia and South Africa. These two countries produce about 85 percent of the world's zirconium.

Isotopes

There are five naturally occurring isotopes of zirconium: zirconium-90, zirconium-91, zirconium-92, zirconium-94, and zirconium-96. Isotopes are two or more forms of an element. Isotopes differ from each other according to their mass number. The number written to the right of the element's name is the mass number. The mass number represents the number of protons plus neutrons in the nucleus of an atom of the element. The number of protons determines the element, but the number of neutrons in the atom of any one element can vary. Each variation is an isotope.

About a dozen radioactive isotopes of zirconium are known also. A radioactive isotope is one that breaks apart and gives off some form of radiation. Radioactive isotopes are produced when very small particles are fired at atoms. These particles stick in the atoms and make them radioactive.

No radioactive isotope of zirconium has any important practical application.

Extraction

Zirconium ores are first converted to zirconium tetrachloride ($ZrCl_4$). This compound is then mixed with magnesium metal at high temperature:

$$2Mg + ZrCl_4 \rightarrow 2MgCl_2 + Zr$$

Uses

Many zirconium alloys are available. They are used to make flash bulbs, rayon spinnerets (the nozzles from which liquid rayon is released), lamp filaments, precision tools, and surgical instruments. These uses make up only a small amount of the

Naturally occurring zircon is in demand as a gemstone. It is polished, cut, and used for jewelry and art.

Zirconium alloys are used to make concrete drill bits.

metal produced, however, compared to its application in nuclear power plants.

Compounds

About 95 percent of all zirconium produced is converted into a compound before being used. The two most common compounds made are zircon (zirconium silicate) and zirconia (zirconium oxide).

Naturally occurring zircon is in demand as a gemstone. It is polished, cut, and used for jewelry and art. Natural zircon often includes **uranium, thorium,** and other radioactive elements. The presence of these elements often gives a zircon a special brilliance and fire-like quality, resembling fine diamonds.

Zircon has other properties that make it desirable in industrial applications. For example, it is an excellent refractory material. A refractory is a material that does not conduct heat well. It is able to withstand very high temperatures without cracking or breaking down.

Zircon is used to make the foundry molds used to make metal pieces of all shapes. Molten metal is poured into the mold. When it cools, it is removed from the mold. The use of zircon in a refractory mold produces a smooth surface on the metal.

Zircon is also used to make bricks in high-temperature furnaces and ovens. These furnaces and ovens are used to work with molten metals. Zircon bricks are ideal for such ovens because they reflect heat and are not destroyed by high temperatures.

Both zircon and zirconia are used as abrasives. An abrasive is a powdery material used to grind or polish other materials. Another important use of zircon and zirconia is in making objects opaque. Opaque means that light is not able to pass through. Suppose a person wants to make a glaze for pottery that looks completely white. The glaze must reflect all light that strikes it and not let any light through. Adding zircon or zirconia to the glaze will achieve this result.

Health effects

Zirconium is regarded as relatively safe. Some studies have shown that it can cause skin irritation, however. Deodorant products containing zirconium have been found to produce skin rashes.

Zirconium can cause skin irritation. Deodorant products containing zirconium have been found to produce skin rashes.

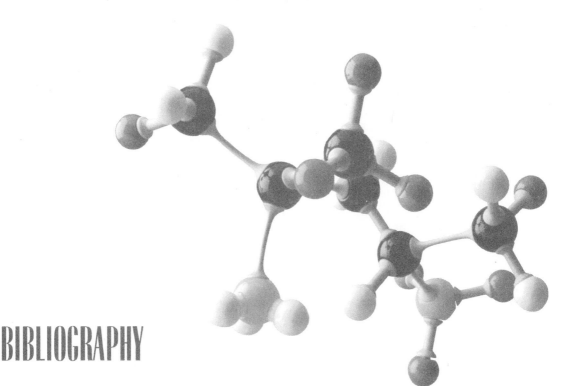

BIBLIOGRAPHY

General sources

Print resources

Atkins, P. W. *The Periodic Kingdom: A Journey into the Land of the Chemical Elements*. New York: HarperCollins, 1997.

Budavari, Susan, ed. *The Merck Index*. Rahway, NJ: Merck & Company, Inc., 1989.

Emsley, John, and Jan Neruda. *Elements*. New York: Oxford University Press, 1996.

Greenwood, N. N., and A. Earnshaw. *Chemistry of the Elements*. Oxford: Pergamon Press, 1984.

Heiserman, David L. *Exploring Chemical Elements and Their Compounds*. New York: TAB Books/McGraw Hill, 1991.

Krebs, Robert E. *The History and Use of Our Earth's Chemical Elements: A Reference Guide*. Westport, CT: Greenwood Publishing Group, 1998.

Lewis, Richard J., Sr. *Hawley's Condensed Chemical Dictionary*, 12th edition. New York: Van Nostrand Reinhold Company, 1993.

Nachaef, N. *The Chemical Elements: The Exciting Story of Their Discovery and of the Great Scientists Who Found Them*. Jersey City, NJ: Parkwest Publications, 1997.

Newton, David E. *The Chemical Elements*. New York: Franklin Watts, 1994.

Ruben, Samuel. *Handbook of the Elements*. Chicago: Open Court Publishing Company, 1985.

Stwertka, Albert. *A Guide to the Elements*. New York: Oxford University Press, 1996.

Trifinov, D. N., and V. D. Trifinov. *Chemical Elements: How They Were Discovered*. New York: State Mutual Books, 1985.

Weeks, Mary Elvira, and Henry M. Leicester. *Discovery of the Elements*, 7th edition. Madison, WI: Journal of Chemical Education, 1968.

Internet sources

Readers should be reminded that some Internet sources change frequently. Some of the following web sites may have been removed and new ones added.

Bentor, Yinon. "The Periodic Table of the Elements on the Internet." http://domains.twave.net/domain/yinon/default.html.

Dayah, Michael. "Periodic Table of the Elements." http://www.benray.com/periodic/.

Hyper Chemistry on the Web. "The Periodic Table of Elements." http://tqd.advanced.org/2690/ptable/ptable.html.

KLB Productions. "Yogi's Behemoth Periodic Table of Elements." http://klbproductions.com/yogi/periodic.

Kostas, Tsigaridis. "Periodic Table of the Elements." http://www.edu.uch.gr/~tsigarid/ptoe/info.html.

Los Alamos National Laboratory. Chemical Division. "Periodic Table of the Elements." http://mwanal.lanl.gov/CST/imagemap/periodic/periodic.html.

Phoenix College. "The Pictorial Periodic Table." http://chemlab.pc.maricopa.edu/periodic/periodic.html.

"Taits Periodic Table of the Elements." http://bvsd.co.edu/~stanglt/per/NN3/index.html.

University of Akron. Department of Chemistry. Hardy Research Group. http://ull.chemistry.uakron.edu/periodic_table/.

Winter, Mark. "The Periodic Table on the WWW: Web Elements." http://www.shef.ac.uk/chemistry/web-elements/web-elements-home.html.

Yahoo. "Periodic Table of the Elements." http://www.yahoo.com/science/chemistry/periodic_table_of_the_elements.

Specific elements

Aluminum

Aluminum Association, Inc., 900 19th Street, N.W., Washington, D.C., 20006-7168; (202) 862-5134. Information available on the Internet at www.aluminum.org.

Americium

"Smoke Detectors and Americium." *Nuclear Issues Briefing Paper 35*. Uranium Information Center Ltd., April 1997. Information available on the Internet at http://www.uic.com.au/nip35.htm.

Yu, Jessen, "A:Eeeeeeeee! There are two different types of smoke detectors—photoelectric detectors and ionization detectors." *The Stanford Daily*, May 7, 1996. Information available on the Internet at http://daily.stanford.edu/5-7-96/NEWS/index.html.

Bismuth

Bismuth Institute, 301 Borgtstraat – B.1850, Grimbergen, Belgium. Information available on the Internet at http://www.bismuth.be/.

Copper

"Copper: The Red Metal." Information available on the Internet at http://spidergram.ccs.unr.edu/unr/sb204/geology/copper2.html.

The Copper Data Center. Information available on the Internet at http://cdc.copper.org/.

Fluorine

Newton, David E. *The Ozone Dilemma*. Santa Barbara, CA: ABC-CLIO, 1995.

Stille, Darlene R. *Ozone Hole*. Chicago: Childrens Press, 1991.

"The Unofficial Polytetrafluoroethylene (PTFE) Homepage." Information available on the Internet at http://www.net-master.net/~ptfedave/.

Gold

Gold Institute, 1112 16th Street, N.W., Suite 240, Washington, D.C., 20036; (202) 835-0185; e-mail at info@goldinstitute.org. The Institute's web page is available at http://www.goldinstitute.com/about.html.

Hydrogen

American Hydrogen Association, 216 S. Clark, #103, Tempe, AZ 85281; (602) 827-7915.

Canadian Hydrogen Association, 8 King's College, Toronto, Ontario, Canada, M8S 1A4; telephone: (416) 978-2531; fax: (416) 978-0787.

National Hydrogen Association, 1800 M Street, Suite 300, Washington, D.C., 20036-5802; telephone: (202) 223-5547; fax: (202) 223-5537.

Indium

Indium Metal Information Center. Information available on the Internet at http://www.indium.com/metalcenter.html.

Iridium

"Bearing down on the kilogram standard." *Science News,* January 28, 1995, p. 63.

Croft, Sally, "Keeping the kilo from gaining weight." *Science,* May 12, 1995, p. 804.

Ralof, Janet, "Unclogging arteries? Radiation helps." *Science News,* June 14, 1997, p. 364.

Iron

American Iron and Steel Institute, 11017 17th Street, N.W., Washington, D.C., 20036. Information available on the Internet at http://www.steel.org.

Lead

International Lead Zinc Research Organization, 2525 Meridian Parkway, P.O. Box 12036, Research Triangle Park, N.C.; telephone: (919) 361-4647. Information available on the Internet at http://www.ilzro.org/ilzro.html.

Lead Industries Association, 292 Madison Ave., New York, NY 10017.

Lithium

Health Center. "Mood Stabilizer Medications: Lithium." Information available on the Internet at http://www.healthguide.com/Pharmacy/LITHIUM.htm.

Molybdenum

International Molybdenum Association, 7 Hackford Walk, 119-123 Hackford Road, London SW9 0QT. Information available on the Internet at http://www.imoa.org.uk/.

Nickel

Nickel Development Institute, 214 King Street West, Suite 510, Toronto, Ontario, Canada; M5H 3S6; telephone: (416) 591-7999; fax: (416) 591-7987. Information available on the Internet at http://www.nidi.org/.

Nickel Producers Environmental Research Association, 2604 Meridian Parkway, Suite 200, Durham, NC 27713; telephone: (919) 544-7722.

Niobium

"What is Niobium?" Information available on the Internet at http://www.teleport.com/~paulec/Niobium.html.

Yarris, Lynn, "Magnet Sets World Record." *Research Review,* fall 1997.

Palladium

"The Amazing Metal Sponge." Information available on the Internet at http://www.psc.edu/MetaCenter/MetaScience/Articles/Wolf/Wolf.html.

Plutonium

"Electronic Resource Library," Amarillo National Resource Center for Plutonium. Information available on the Internet at http://plutonium-erl.actx.edu/.

Radium

Birch, Beverly. *Marie Curie.* Milwaukee, WI: Gareth Stevens Publishing, 1988.

Keller, Mollie. *Marie Curie.* New York: Franklin Watts, 1982.

Parker, Steve. *Marie Curie and Radium.* New York: HarperCollins, 1992.

Pflaum, Rosalynd. *Grand Obsession: Madame Curie and Her World.* New York: Doubleday, 1989.

Poynter, Margaret. *Marie Curie: Discoverer of Uranium.* Hillside, NJ: Enslow Publishers, 1994.

Saari, Peggy, and Stephen Allison, eds. *Scientists: The Lives and Works of 150 Scientists.* Detroit: U•X•L, 1996, pp. 181–91.

Radon

"A Citizen's Guide to Radon: The Guide to Protecting Yourself and Your Family from Radon." Washington, D.C.: U.S. Environmental Protection Agency, 1992.

Selenium

Selenium-Tellurium Development Association, 11 Broadway, New York, NY 10013. Information available on the Internet at http://www.stda.be/.

Sulfur

Sulfur Institute, 1140 Connecticut Avenue, N.W., Washington, D.C., 20036.

Tellurium

Selenium-Tellurium Development Association, 11 Broadway, New York, NY 10013. Information available on the Internet at http://www.stda.be/.

Tin

Tin Information Center of North America, 1353 Perry Avenue, Columbus, OH 43201.

U.S. Geological Survey, *Minerals Information-1996,* Washington, D.C.: Government Printing Office, 1997.

Uranium

"The Core," The Uranium Institute. Information available on the Internet at http://www.uilondon.org.

Uranium & Nuclear Power Information Centre, Australia. Information available on the Internet at http://www.uic.com.au/index.htm.

Vanadium

Hilliard, Henry E. "Vanadium," *Minerals Yearbook.* Washington, D.C.: U.S. Geological Survey.

McGuire, Rory, "Vanadium battery research recharged by Pinnacle Mining." *Uniken,* June 27, 1997, pp. 6–10. Information available on the Internet at http://www.ceic.unsw.edu.au/centers/vrb/Pinnacle.htm.

Zinc

American Zinc Association, 1112 16th Street, N.W., Washington, D.C., 20036.

International Lead Zinc Research Organization, 2525 Meridian Parkway, P.O. Box 12036, Research Triangle Park, N.C.; telephone: (919) 361-4647. Information available on the Internet at http://www.ilzro.org/ilzro.html.

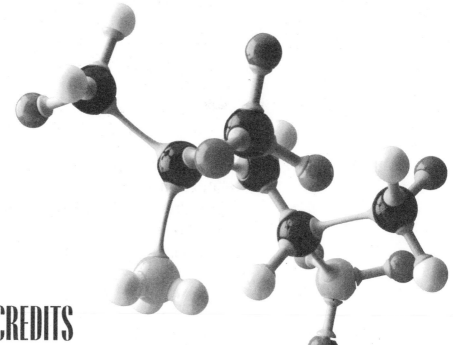

PICTURE CREDITS

The photographs appearing in *Chemical Elements: From Carbon to Krypton* were reproduced by permission of the following sources:

On the front cover: **Kenneth Eward / BioGrafx, National Audubon Society Collection / Photo Researchers, Inc.:** background image; **Yoav Levy / Phototake NYC:** first box; **Andrew Syred / The National Audubon Society Collection / Photo Researchers, Inc.:** third box.

On the back cover: **Kenneth Eward / BioGrafx, National Audubon Society Collection / Photo Researchers, Inc.:** first box; **Yoav Levy / Phototake NYC:** second box; **Lawrence Migdale / Science Source, National Audubon Society Collection / Photo Researchers, Inc.:** third box.

In the text: **JLM Visuals:** pp. 10, 91, 93, 151, 271, 548, 561; **Library of Congress:** pp. 11, 35, 88, 104, 121, 127, 158, 167, 180, 211, 364, 398, 435, 467, 468, 472, 474, 479, 493; **Field Mark Publications:** pp. 12, 17, 23, 34, 85, 116, 126, 154, 164, 189, 198, 228, 256, 264, 265, 322, 323,

338, 352, 387, 393, 412, 420, 421, 448, 489, 522, 537, 550, 583, 602, 619, 651, 655, 678, 685; **Russ Lappa / Science Source, National Audubon Society Collection / Photo Researchers, Inc.:** pp. 22, 143, 203, 227, 258, 369, 377, 427, 497, 581, 611, 635, 649, 683; **Alexander Tsiaras / National Audubon Society Collection / Photo Researchers, Inc.:** p. 28; **Michael English / Custom Medical Stock Photo, Inc.:** p. 43; **Luis de la Maza / Phototake NYC:** p. 44; **Corbis-Bettmann:** pp. 50, 252, 278, 283, 450, 629; **Mark Elias / AP/Wide World Photos, Inc.:** p. 56; **Earl Scott / National Audubon Society Collection / Photo Researchers, Inc.:** p. 62; **AP/Wide World Photos, Inc.:** pp. 69, 98, 403; **Ray Nelson / Phototake NYC:** p. 70; **Charles D. Winters / National Audubon Society Collection / Photo Researchers, Inc.:** pp. 75, 109, 309, 418, 527, 667; **Department of Energy:** p. 78; **Rich Treptow / National Audubon Society Collection / Photo Researchers, Inc.:** pp. 83, 519; **University of California-Berkeley:** p. 99; **Andrew Mcclenaghan / National Audubon Society Collection / Photo Researchers, Inc.,** p. 105; **Hans Namuth / Photo Researchers, Inc.:** p. 110; **Royal Greenwich Observatory / National Audubon Society Collection / Photo Researchers, Inc.:** p. 124; **Yoav Levy / Phototake NYC:** pp. 129, 140, 320, 353, 665; **Bruce Iverson / Science Photo Library, National Audubon Society Collection/ Photo Researchers, Inc.:** p. 144; **New York Convention & Visitors Bureau, Inc.:** p. 152; **Roland Seitre / BIOS:** p. 156; **U.S. National Aeronautics and Space Administration (NASA):** pp. 160, 404, 441, 607; **Will and Deni McIntrye / National Audubon Society Collection / Photo Researchers, Inc.:** pp. 171, 656; **Erich Schrempp / National Audubon Society Collection / Photo Researchers, Inc.:** p. 172; **Visual Image:** p. 184; **Science Photo Library:** p. 193; **Alfred Pasieka / Science Photo Library, The National Audubon Society Collection / Photo Researchers, Inc.:** pp. 205, 263, 281; **Michael Abbey / Science Source, The National Audubon**

ety Collection / Photo Researchers, Inc.: p. 475; **Hank Morgan / National Audubon Society Collection / Photo Researchers, Inc.:** p. 482; **David Sutton / Zuma Images / The Stock Market:** p. 504; **Red Elf / Denise Ward-Brown:** p. 507; **Trek Bicycle Corp.:** p. 515; **Adam Hart-Davis / Photo Researchers, Inc.:** p. 530; **Andrew Syred / The National Audubon Society Collection / Photo Researchers, Inc.:** p. 531; **Bryan Peterson / Stock Market:** p. 535; **Lawrence Migdale / Science Source, National Audubon Society Collection / Photo Researchers, Inc.:** pp. 544, 562; **Fireworks by Grucci:** p. 557; **David Taylor / Science Photo Library, National Audubon Society Collection / Photo Researchers, Inc.:** p. 566; **Custom Medical Stock Photo, Inc.:** p. 572; **National Audubon Society Collection / Photo Researchers, Inc.:** 588; **Peter Berndt, M.D. / Custom Medical Stock Photo:** p. 594; **Tony Freeman / PhotoEdit:** p. 600; **Account Phototake / Phototake NYC:** pp. 620, 622; **National Oceanic & Atmospheric Administration:** p. 637; **Paolo Koch / National Audubon Society Collection / Photo Researchers, Inc.:** p. 641; **Department of Energy:** p. 642; **PHOTRI / Stock Market:** p. 643; **Robert L. Wolke:** p. 644; **Rick Altman / The Stock Market:** p. 675; **Archive fur Kunst und Geschichte / Photo Researchers, Inc.:** p. 677.

In the color signatures, volume 1: **Erich Schrempp / National Audubon Society Collection / Photo Researchers, Inc.:** p. 1 (top); **Andrew Mcclenaghan / National Audubon Society collection / Photo Researchers, Inc.:** p. 1 (top); **Charles D. Winters / National Audubon Society Collection / Photo Researchers, Inc.:** pp. 2 (top), 3 (bottom); **Yoav Levy / Phototake NYC:** pp. 2 (bottom), 7 (top), 8; **JLM Visuals:** p. 3 (top); **Lawrence Berkeley Lab / National Audubon Society Collection / Photo Researchers, Inc.:** pp. 4–5; **Earl Scott / National Audubon Society Collection / Photo Researchers, Inc.:** p. 6 (top); **Alexander Tsiaras / National Audubon Society Collection / Photo Researchers, Inc.:** p. 6 (bottom); **Rich Treptow / Nation-**

al Audubon Society Collection / Photo Researchers, Inc.: p. 7 (bottom).

In the color signatures, volume 2: **Chris Jones / The Stock Market:** p. 1 (top); **Philippe Plailly / Science Photo Library, The National Audubon Society Collection / Photo Researchers, Inc.:** p. 1 (bottom); **Chris Rogers / The Stock Market:** p. 2 (top); **Michael W. Davidson / National Audubon Society Collection / Photo Researchers, Inc.:** p. 2 (bottom); **Dornier Space/Science Photo Library, National Audubon Society Collection / Photo Researchers, Inc.:** p. 3 (top); **Robert Visser / Greenpeace:** p. 3 (bottom); **Tom Ives / The Stock Market:** pp. 4–5; **Ronald Royer / Science Photo Library, National Audubon Society Collection / Photo Researchers, Inc.:** p. 6; **Russ Lappa / Science Source, National Audubon Society Collection / Photo Researchers, Inc.:** p. 7 (top); **Thomas Del Brase / Stock Market:** p. 7 (bottom); **Yoav Levy / Phototake NYC:** p. 8.

In the color signatures, volume 3: **JLM Visuals:** p. 1 (top); **Science Source / National Audubon Society Collection / Photo Researchers, Inc.:** p. 1 (bottom); **Scott Camazine / National Audubon Society Collection / Photo Researchers, Inc.:** p. 2; **Department of Energy:** p. 3 (top); **U.S. National Aeronautics and Space Administration (NASA):** p. 3 (bottom); **PHOTRI / Stock Market:** pp. 4–5; **Lawrence Migdale / Science Source, National Audubon Society Collection / Photo Researchers, Inc.:** p. 6 (top); **Bryan Peterson / Stock Market:** p. 6 (bottom); **Charles D. Winters / National Audubon Society Collection / Photo Researchers, Inc.:** p. 7 (top); **Tony Ward / Photo Researchers, Inc.:** p. 7 (bottom); **National Audubon Society Collection / Photo Researchers, Inc.:** p. 8.

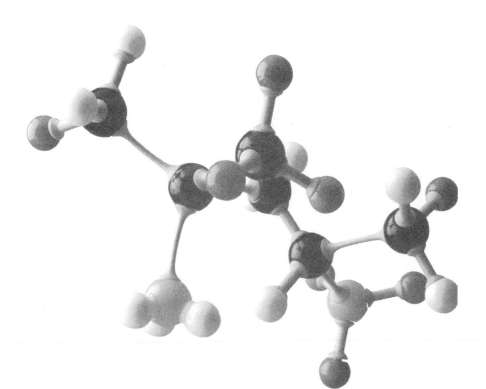

INDEX

A

Abel's metal, *1:* 82

Abelson, Philip H., *2:* 363; *3:* 433

Abrasives, *1:* 71, 117; *3:* 686

Absolute zero, *3:* 666

Ac. *See* Actinium (Ac)

Acetylene gas, *2:* 405

Acids, *2:* 253–54

Actinium (Ac), *1:* **1–3,** 193; *3:* 467

Actinides. *See* Americium (Am); Berkelium (Bk); Californium (Cf); Curium (Cm); Einsteinium (Es); Fermium (Fm); Lawrencium (Lr); Mendelevium (Md); Neptunium (Np); Nobelium (No); Plutonium (Pu); Protactinium (Pa); Thorium (Th); Uranium (U)

Adenosine triphosphate (ATP), *3:* 422

Aerosol cans, *1:* 189 (ill.)

Ag. *See* Silver (Ag)

Agricola, Georgius, *1:* 185

Air, *2:* 382, 396

Airline baggage, *1:* 100

Airport runway lights, *2:* 292

Al. *See* Aluminum (Al)

Alabamine, *1:* 38, 192

Alaska, elements in, *2:* 212, 221, 272, 345; *3:* 674

Albania, elements in, *1:* 137

Albert the Great (Albertus Magnus), *1:* 32

Alchemy, *1:* 19, 20, 31; *3:* 415, 672

Alcohols, *1:* 111

Algae, *3:* 423

Algeria, elements in, *2:* 336

Alka-Seltzer, *3:* 549, 550 (ill.)

Alkali metals. *See* Cesium (Cs); Francium (Fr); Lithium (Li); Potassium (K); Rubidium (Rb); Sodium (Na)

Alkaline earth metals. *See* Barium (Ba); Beryllium (Be); Calcium (Ca); Magnesium (Mg); Radium (Ra); Strontium (Sr)

Alkalis, *3:* 427

Allanite, *2:* 352

Allergies, *2:* 374

Allison, Fred, *1:* 38, 191–92

Alloys. *See* under "Uses" category in most entries

Alpha particles, *2:* 234–35

Alum, *1:* 6

Aluminosis, *1:* 13

Aluminium. *See* Aluminum

Aluminum (Al), *1:* **5–13,** 10 (ill.), 11 (ill.), 12 (ill.), 54; *2:* 310; *3:* 614

Italic type indicates volume number; **boldface** indicates main entries and their page numbers; (ill.) indicates photos and illustrations.

CHEMICAL **elements**

N

CHEMICAL **elements**

Main–Group Elements

Atomic number 86   (222)   **Atomic weight**
Symbol Rn
Name radon

Transition Metals

Period	1 IA	2 IIA	3 IIIB	4 IVB	5 VB	6 VIB	7 VIIB	8	9 VIIIB
1	1 1.00794 H hydrogen								
2	3 6.941 Li lithium	4 9.012182 Be beryllium							
3	11 22.989768 Na sodium	12 24.3050 Mg magnesium							
4	19 39.0983 K potassium	20 40.078 Ca calcium	21 44.955910 Sc scandium	22 47.88 Ti titanium	23 50.9415 V vanadium	24 51.9961 Cr chromium	25 54.9305 Mn manganese	26 55.847 Fe iron	27 58.93320 Co cobalt
5	37 85.4678 Rb rubidium	38 87.62 Sr strontium	39 88.90585 Y yttrium	40 91.224 Zr zirconium	41 92.90638 Nb niobium	42 95.94 Mo molybdenum	43 (98) Tc technetium	44 101.07 Ru ruthenium	45 102.90550 Rh rhodium
6	55 132.90543 Cs cesium	56 137.327 Ba barium	57 138.9055 *La lanthanum	72 178.49 Hf hafnium	73 180.9479 Ta tantalum	74 183.85 W tungsten	75 186.207 Re rhenium	76 190.2 Os osmium	77 192.22 Ir iridium
7	87 (223) Fr francium	88 (226) Ra radium	89 (227) †Ac actinium	104 (261) Rf rutherfordium	105 (262) Db dubnium	106 (263) Sg seaborgium	107 (262) Bh bohrium	108 (265) Hs hassium	109 (267) Mt meitnerium

*Lanthanides

58 140.115 Ce cerium	59 140.90765 Pr praseodymium	60 144.24 Nd neodymium	61 (145) Pm promethium	62 150.36 Sm samarium

†Actinides

90 232.0381 Th thorium	91 (231) Pa protactinium	92 238.0289 U uranium	93 (237) Np neptunium	94 (244) Pu plutonium